L. E Brown

The Contractors', Builders', and Carpenters' Building Table and Estimate Book

Being a systematic, thorough and practical method by which to estimate the cost

L. E Brown

The Contractors', Builders', and Carpenters' Building Table and Estimate Book
Being a systematic, thorough and practical method by which to estimate the cost

ISBN/EAN: 9783337070939

Printed in Europe, USA, Canada, Australia, Japan

Cover: Foto ©Suzi / pixelio.de

More available books at **www.hansebooks.com**

THE
CONTRACTORS', BUILDERS',

AND

CARPENTERS'
Building Table and Estimate Book,

— BEING —

A SYSTEMATIC, THOROUGH AND PRACTICAL METHOD
BY WHICH TO ESTIMATE THE COST OF BUILDING
WORK AND MATERIAL, WITH FULL AND
COMPLETE INSTRUCTIONS.

TOGETHER WITH A GREAT VARIETY OF FULL BILLS OF TIMBER,
GIVING THE NUMBER OF FEET OF FRAME LUMBER CONTAINED
IN OVER TWO HUNDRED BUILDINGS OF VARIOUS DIMEN-
SIONS; ALSO THE NUMBER OF FEET OF BOARDING,
CLAPBOARDS AND SHINGLES REQUIRED TO COV-
ER THE SAME, WITH MANY OTHER VALUA-
BLE TABLES FOR BUILDING PIAZZAS,
BAY WINDOWS, PORTICOS, &C.,

BY

L. E. BROWN,

BUILDER.

WINCHENDON:
PUBLISHED BY THE AUTHOR.

1878.

Entered according to Act of Congress, in the year 1878. by
L. E. BROWN,
in the Office of the Librarian of Congress at Washington, D. C,

CONTENTS.

Page.
9 to 12, explanation of bills of timber.

BALLOON FRAMES.

Page.
12, Building Nos. 1 to 3, 8 by 10 feet—three different heights.
13 " "   4 to 6, 10 by 12 feet— "   "   "
14 " "   7 to 9, 10 by 14 feet— "   "   "
15 " "   10 to 12, 10 by 16 feet— "   "   "
16 " "   13 to 15, 12 by 12 feet— "   "   "
17 " "   16 to 18, 12 by 14 feet— "   "   "
18 " "   19 to 21, 12 by 16 feet— "   "   "
19 " "   22 to 24, 14 by 14 feet— "   "   "
20 " "   25 to 27, 14 by 16 feet— "   "   "
21 " "   28 to 30, 14 by 18 feet— "   "   "
22 " "   31 to 33, 14 by 20 feet— "   "   "
23 " "   34 to 36, 16 by 16 feet— "   "   "
24 " "   37 to 39, 16 by 18 feet— "   "   "
25 " "   40 to 42, 16 by 20 feet— "   "   "
26 " "   43 to 45, 16 by 22 feet— "   "   "
27 " "   46 to 48, 16 by 24 feet— "   "   "
28 " "   49 to 51, 18 by 18 feet— "   "   "
29 " "   52 to 54, 18 by 20 feet— "   "   "
30 " "   55 to 57, 18 by 22 feet— "   "   "
31 " "   58 to 60, 18 by 24 feet— "   "   "
32 " "   61 to 63, 18 by 26 feet— "   "   "
33 " "   64 to 66, 18 by 28 feet— "   "   "
34 " "   67 to 69, 18 by 30 feet— "   "   "
35 " "   70 to 72, 20 by 20 feet— "   "   "
36 " "   73 to 75, 20 by 22 feet— "   "   "
37 " "   76 to 77, 20 by 24 feet— 2 different heights.
38 " "   78, 20 by 24 feet—20 feet posts.
38 " "   79, 20 by 26 feet—12   "
39 " "   80, 20 by 26 feet—14   "
39 " "   81, 20 by 26 feet—20   "
40 " "   82, 20 by 28 feet—12   "
40 " "   93, 20 by 28 feet—14   "
41 " "   84, 20 by 28 feet—20   "
41 " "   85, 20 by 30 feet—12   "
42 " "   86, 20 by 30 feet—14   "
42 " "   87, 20 by 30 feet—20   "
43 " "   88, 22 by 22 feet—22   "
43 " "   89, 22 by 22 feet—14   "

Page.			
44,	Building No.	90, 22 by 22 feet—20	"
44	"	" 91, 22 by 24 feet—12	"
45	"	" 92, 22 by 24 feet—14	"
45	"	" 93, 22 by 24 feet—20	"
46	"	" 94, 22 by 26 feet—12	"
46	"	" 95, 22 by 26 feet—14	"
46	"	" 96, 22 by 26 feet—20	"
47	"	" 97, 22 by 28 feet—12	"
47	"	" 98, 22 by 28 feet—14	"
48	"	" 99, 22 by 28 feet—20	"
48	"	" 100, 22 by 30 feet—12	"
49	"	" 101, 22 by 30 feet—14	"
49	"	" 102, 22 by 30 feet—20	"

ONE-HALF BALLOON FRAMES.

49,	Building No.	103, 24 by 24 feet—12 feet posts.	
50,	"	" 104, 24 by 24 feet—14	"
50	"	" 105, 24 by 24 feet—20	"
51	"	" 106, 24 by 26 feet—12	"
51	"	" 107, 24 by 26 feet—14	"
52	"	" 108, 24 by 26 feet—20	"
52	"	" 109, 24 by 28 feet—12	"
53	"	" 110, 24 by 28 feet—14	"
53	"	" 111, 24 by 28 feet—20	"
54	"	" 112, 24 by 30 feet—12	"
54	"	" 113, 24 by 30 feet—14	"
55	"	" 114, 24 by 30 feet—20	"
55	"	" 115, 26 by 26 feet—12	"
56	"	" 116, 26 by 26 feet—14	"
56	"	" 117, 26 by 20 feet—20	"
57	"	" 118, 26 by 28 feet—12	"
57	"	" 119, 26 by 28 feet—14	"
58	"	" 120, 26 by 28 feet—20	"
58	"	" 121, 26 by 30 feet—12	"
59	"	" 122, 26 by 30 feet—14	"
59	"	" 123, 26 by 30 feet—20	"
60	"	" 124, 26 by 32 feet—12	"
60	"	" 125, 26 by 32 feet—14	"
61	"	" 126, 26 by 32 feet—20	"
61	"	" 127, 26 by 34 feet—12	"
62	"	" 128, 26 by 34 feet—14	"

CONTENTS.

Page.		
62,	Building No. 129,	26 by 34 feet—20 feet posts.
63,	"	130, 26 by 36 feet—12 "
	"	131, 26 by 36 feet—14 "
64,	"	132, 26 by 36 feet—20 "
	"	133, 26 by 38 feet—12 "
65,	"	134, 26 by 38 feet 14 "
	"	135, 26 by 38 feet—20 "
66	"	136, 26 by 40 feet—12 "
	"	137, 26 by 40 feet—14 "
67	"	138, 26 by 40 feet—20 "
68	"	139, 28 by 28 feet—12 "
	"	140, 28 by 28 feet—14 "
69	"	141, 28 by 28 feet—20 "
	"	142, 28 by 30 feet—12 "
70	"	143, 28 by 30 feet—14 "
	"	144, 28 by 30 feet—20 "
71	"	145, 28 by 32 feet—12 "
	"	146, 28 by 32 feet—14 "
72	"	147, 28 by 32 feet—20 "
73	"	148, 28 by 34 feet—16 "
	"	149, 28 by 34 feet—20 "
74	"	150, 28 by 34 feet—12 "

FULL FRAMES.

75	Building No.	151, 30 by 30 feet—18 "
76	"	152, 30 by 30 feet—22 "
	"	153, 30 by 30 feet—12 "
77	"	154, 30 by 32 feet—18 "
78	"	155, 30 by 32 feet—22 "
	"	156, 30 by 32 feet—12 "
79	"	157, 30 by 34 feet—18 "
80	"	158, 30 by 34 feet—22 "
	"	159, 30 by 34 feet—12 "
81	"	160, 30 by 36 feet—18 "
	"	161, 30 by 36 feet—22 "
82	"	162, 30 by 36 feet—12 "
	"	163, 32 by 32 feet—18 "
83	"	164, 32 by 32 feet—22 "
	"	165, 32 by 32 feet—12 "

CONTENTS.

Page.			
83,	Building	No. 166, 32 by 34 feet—18	feet posts.
84	"	" 167, 32 by 34 feet—22	"
84	"	" 168, 32 by 34 feet—12	"
85	"	" 169, 32 by 36 feet—18	"
85	"	" 170, 32 by 36 feet—22	"
86	"	" 171, 32 by 36 feet—12	"
86	"	" 172, 32 by 38 feet—18	"
87	"	" 173, 32 by 38 feet—22	"
87	"	" 174, 32 by 38 feet—12	"
88	"	" 175, 34 by 34 feet—18	"
88	"	" 176, 34 by 34 feet—22	"
89	"	" 177, 34 by 34 feet—12	"
89	"	" 178, 34 by 36 feet—18	"
90	"	" 179, 34 by 36 feet—22	"
90	"	" 180, 34 by 36 feet—12	"
90	"	" 181, 34 by 38 feet—18	"
91	"	" 182, 34 by 38 feet—22	"
91	"	" 183, 34 by 38 feet—12	"
92	"	" 184, 36 by 40 feet—18	"
92	"	" 185, 34 by 40 feet—22	"
93	"	186, 34 by 40 feet—12	"
	"	187, 30 by 40 feet—18	"
94	"	193, 36 by 40 feet—18	"
	"	194, 36 by 40 feet—22	"
95	"	191, 32 by 40 feet—22	"
	"	192, 32 by 40 feet—12	"
96	"	188, 30 by 40 feet—22	"
	"	189, 30 by 40 feet—12	"
	"	190, 32 by 40 feet—18	"
97	"	195, 36 by 40 feet—12	"
	"	196, 38 by 40 feet—18	"
98	"	197, 38 by 40 feet—22	"
	"	198, 38 by 40 feet—12	"
99	"	199, 40 by 40 feet—18	"
	"	200, 40 by 40 feet—22	"
100	"	201, 40 by 40 feet—12	"
	"	202, 40 by 60 feet—18	"
101	"	203, 40 by 50 feet—22	"
	"	204, 40 by 60 feet—18	"

Page.		
102	Building No. 205, 40 by 60 feet—22 feet posts.	
102	" 206, 45 by 60 feet—18	"
103	" 207, 45 by 60 feet—22	"

BARN FRAMES.

103	" 208, 26 by 30 feet—16	"
104	" 209, 30 by 36 feet—16	"
105	" 210, 30 by 40 feet—16	"
106	" 211, 36 by 40 feet—16	"
107	" 212, 40 by 40 feet—16	"
108	" 213, 40 by 50 feet—16	"
109	" 214, 40 by 60 feet—16	"
110	" 215, 40 by 70 feet—16	"
111	" 216, 45 by 84 feet—16	"
112	" 217, 50 by 100 feet—16	"
114	Contractors' and Builders' Estimate.	
115	Close Estimate.	
116	Carpenter work:	
117	Window Frames Casings and Door Frames.	
118	Door Casings, Door Frames, Stairs, &c.	
119	Window and Door Frames cased.	
120	Jets.	
121	Table of Nails.	
122	Sheathing, Flooring, Boarding, and Top Floors.	
123	Clapboards, Laths, Shingles, Sand, Lime, &c.	
124	Bricks, Chimneys, Cement for Brick, &c.	
125	Lead Pipe, Furring, Partition Studs and Wall studding	
126	Piazzas.	
127	Piazzas, Bay Windows and Porticos.	
128	Table of Joists and Studding.	
129	" " " "	
130	" " " "	
131	" " " "	
132	Rafter Table.	
133	Braces.	
134	Board Measure, Square Timber and joists.	
135	Square timber concluded.	
136	Hay, Iron, and Measuring Wood and Bark.	
137	Measuring Wood and Bark.	
138	Measurement of Wood and Bark, continued.	
139	Hip Rafters.	
140	Octogons.	
141	Painting and Mansard Roof.	
142 to 145	Pitch of Roofs.	

PREFACE.

The Author does not claim any invention or new discovery; there is nothing contained in the work that should not be familiar to all experienced builders. But to such, as a book of reference, it will be of great convenience, saving them much time and labor in making estimates. Many builders have long experienced the need of such a work, that would greatly lessen the time and labor in estimating the cost of building, and at the same time teach the workman the art of calculating the cost of buildings. The special object of which is to show the proper method of obtaining the cost of any contract, large or small, and to greatly lessen the time and labor in making the calculations. The greater part of the book is unlike any other work, with so great a variety of full bills of timber giving the number of feet of frame lumber, boarding, clapboarding, shingles &c., in over two hundred buildings of various dimensions, commencing with the size of 8 by 10 and gradually increasing to 100 feet in length with proportionate width. With such a variety one can hardly fail of finding the desired size with the number of feet annexed The rules for measuring dimension lumber, wood and bark, will here be found the shortest and best methods for making such measurements, giving such examples and explanations as *cannot* be found in any Arithmetic. These rules and examples are so thoroughly explained in their several different ways that they should be in the hands of every man and every schoolboy.

Winchendon, Mass., Oct. 1878.   L. E. BROWN.

EXPLANATION OF THE

BILLS OF TIMBER.

There is one general rule that builders use alike in all parts of the country for framing buildings of ordinary sizes. Hence, in estimating the amount of frame lumber contained in a building, this rule is applied: All studding and floor joists are placed 16 inches from centre to centre and all rafters 24 inches from centers for a shingled roof; if a roof is to be slated, place the rafters 16 inches apart, same distance as floor joists and studding. Every bill of lumber contained in this book is estimated by this rule: All studding and floor joists are placed 16 inches from centre to centre, and all rafters 24 inches from centers. In making out a full bill of lumber for buildings of so great a variety and dimensions the author does not claim that all cross timbers in each bill of timber is placed to correspond with the wants of every building of such dimensions, as such timbers vary in their places as the rooms may vary in size. But it must be borne in mind that such alterations of cross timbers does not vary the number of feet of lumber in the building, unless the rooms are made unusually small as to require an extra stick, which I do not think will occur in one instance. The variations of such cross timbers simply alters the length of some of the floor timber. floor joists, girts, &c. when the number feet of lumber remains the same; but in many cases you will find the exact bill without any such alterations.

In making out a bill of timber for a building of much size, there are no two men that would make the same allowance for breakage and waste of studding and joists. No practical man will deny this fact; in a building 25 feet square, it will take 75 studs to go round it, making no allowance for the posts, and 19 studs for one end of the gable in length to reach to the peak of the roof from the centre of the building; it is not safe to allow that these 19 gable studs will stud both gable ends, as there is more waste in cutting; and there is studding wanted for shoring and furrowing, for jetting for gables &c., as I will explain more fully in another place; hence it cannot be expeted that any two practical men will make precisely the same allowance.

In making out the bills of timber contained in this book, each having been carefully estimated, and in all cases great care has been taken to make all proper allowance for waste, &c., that there should be no deficiences.

WHAT IS INCLUDED IN THE BILLS OF LUMBER.

The bills explain; nothing is included that is not billed. All rough lumber is included except the inside partition studs, stair brackets and furrowing. These three items could not be included and be practical, as buildings of the same dimensions may require three times the number of partition studs as another building of the same size, varying according to the number of partitions. Hence, by omitting these three items, a practical estimate can be given, as according to the rules of framing alike in all parts of the country for buildings of ordinary dimensions, studding and floor joists being placed a given distance apart, 16 inches from centers; therefore being governed by this rule it must take a certain amount of lumber and no more, to frame and raise a building of a given size; the size of lumber being given in each estimate.

In giving the length of the rafters a proper allowance has been made for the projection. All rafters are estimated nine inches rise to the foot; therefore all shingles in the bills of estimate, are estimated at that pitch of roof. When any other pitch of a roof may be required, the length of the rafters may be found in the rafter table, in such cases strike out the amount of lumber in the roof and make a new estimate of the same, which will require but a few moments work.

In estimating the number of feet of enclosing, lining and top floor boards to cover the buildings that are estimated in this book, each one has been estimated solid—as if there were to be no doors and windows—and that number of feet set down; if square edge boards are used, add 10 per cent.; if matched add 20 to 30 per cent., according to the width used.

Clapboards also are estimated the same as the boarding, paying no regard to doors and windows. Add one-fourth to that number of feet for buildings of ordinary number of doors and windows, and the estimate will be very near correct for spruce clapboards. which require a little more than one-fourth lap to make good work. The extra lap and waist will about make up for the doors and windows. Mention will be made of clapboards in another place.

The special object of the bills of lumber is to greatly lessen the time and labor for young or old contractors in making an estimate of the amount of lumber in a building of such dimensions as may be found in this book; the variety being so great and the increase in size so gradual, that one can hardly fail of finding the desired size. The author for many years a practical builder and contractor, has often felt the need of a work that would save the labor of many calculations. Hoping that the work will prove of practical value to others besides the author it is submitted to the judgement of the carpenters and builders of the country.

BALLOON FRAMES.

No. 1 building 8 by 10 feet, posts 8 feet in height.

				Feet.
2 sills,	6 x 6 inches,	10 feet long	60
1 "	6 " 6 "	8 " "	48
6 floor joists,	2 " 6 "	10 " "	60
4 posts,	4 " 4 "	8 " "	43
35 studs,	2 " 4 "	8 " "	186
12 rafters,	2 " 5 "	6 " "	60

Total number of feet, ——— 457

Enclosing and under floor boards, 536 feet.
Top floor boards, 80 "
Shingles, 1⅛ M. Clapboards, 390 feet.

No. 2 building, 8 by 10 feet, posts 10 feet in height.

Same as No. 1, with the additional length of posts and studding

Total number of feet, 515.

Enclosing and under floor boards, 608 feet.
Top floor boards, 80 "
Shingles 1⅛ M. Clapboards, 475 feet.

No. 3 building. 8 by 10 feet, posts 12 feet in height

				feet.
2 sills,	6 x 6 inches,	10 feet long,	60
2 "	6 x 6 "	8 "	48
6 1st floor joist	2 x 6 "	10 "	60
7 2d. "	2 x 6 "	10 "	70
4 posts,	4 x 4 "	12 "	64
52 studs,	2 x 4 "	12 "	256
12 rafters,	2 x 5 "	6 "	60

Total number of feet, ——— 618

Enclosing and under floor boards. 750 feet.
Top floor boards, 160 "
Shingles 1⅛ M. Clapboards, 570 feet.

BALLOON FRAMES.

No. 4 building, 10 by 12 feet, posts 8 feet in height.

			feet.
2 sills,	6 x 6 inches,	12 feet long,	72
2 sills,	6 x 6 "	10 "	60
8 floor joists,	2 x 6 "	12 "	96
4 posts,	4 x 4 "	8 "	43
45 studs,	2 x 4 "	8 "	240
14 rafters,	2 x 5 "	7 "	83
	Total number of feet,	——	594

Enclosing and under floor boards, 708 feet

Top floor boards, 120 feet.

Shingles, 1½ M. Clapboards, 490 feet.

No. 5 building, 10 by 12 feet, posts 10 feet in height.

Same as No. 4, with the additional length of posts and studding.

Total number of feet, 664

Enclosing and under floor boards, 796 feet.

Top floor boards, 120 feet.

Shingles, 1½ M. Clapboards, 545 feet.

No. 6 building, 10 by 12 feet, posts 12 feet in height.

			feet.
2 sills,	6 x 6 inches,	12 feet long,	72
2 sills,	6 x 6 "	10 "	60
8 1st floor joists	2 x 6 "	12 "	96
9 2d. "	2 x 6 "	12 "	108
4 posts,	4 x 4 "	12 "	64
42 studs,	2 x 4 "	12 "	256
14 rafters,	2 x 5 "	7 "	83
	Total number of feet,	———	739

Enclosing and under floor boards, 1094 feet.

Top floor boards 240 feet.

Shingles, 1½ M. Clapboards, 600 feet.

BALLOON FRAMES.

No. 7 building, 10 by 14 feet, posts 8 feet in height.

		feet.
2 sills,	6 x 6 inches, 14 feet long,	84
2 "	6 x 6 " 10 "	60
10 floor joists,	2 x 6 " 10 "	100
4 posts,	4 x 4 " 8 "	43
48 studs.	2 x 4 " 8 "	256
16 rafters,	2 x 5 " 7 "	93
	Total number of feet,	—— 636

Enclosing and under floor boards, 788 feet.
Top floor boards, 140 feet.
Shingles, 1¾ M. Clapboards, 530 feet.

No. 8. building, 10 by 14 feet, posts 10 feet in height. Same as No. 7 with the additional length of posts and studding.

Total number of feet, 716

Enclosing and under floor boards, 884 feet.
Top floor boards, 140 feet.
Shingles, 1¾ M. Clapboards, 650 feet.

No. 9 building, 10 by 14 feet, posts 12 feet in height.

		feet.
2 sills,	6 x 6 inches, 14 feet long,	84
2 "	6 x 6 " 10 "	60
10 1st floor joists	2 x 6 " 10 "	100
11 2d. "	2 x 6 " 10 "	110
4 posts,	4 x 4 " 12 "	64
45 studs,	2 x 4 " 12 "	360
16 rafters,	2 x 5 " 7 "	93
	Total number of feet,	—— 871

Enclosing and under floor boards, 1120 feet.
Top floor boards, 280 feet.
Shingles, 1¾ M. Clapboards, 770 feet.

BALLOON FRAMES.

No. 10 building, 10 by 16 feet, posts 8 feet in height.

                                                                                 Feet.

2 sills,	6 x 6 inches,	16 feet long,		96
2 sills,	6 x 6 "	10 "		60
12 floor joists,	2 x 6 "	10 "		120
4 posts,	4 x 4 "	8 "		43
52 studs,	2 x 4 "	8 "		277
18 rafters,	2 x 5 "	7 "		105
		Total number of feet,	———	701

Enclosing and under floor boards, 868 feet,
           Top floor boards, 160 feet.
Shingles, 2 M. Clapboards, 470 feet.

No. 11 building, 10 by 16 feet, posts 10 feet in height.

    Same as No. 10, with the additional length of posts and studding.

                        Total number of feet,   780

Enclosing and under floor boards, 975 feet.
           Top floor boards, 160 feet.
Shingles, 2 M. Clapboards, 700 feet.

No. 12 building, 10 by 16 feet, posts 12 feet in height.

                                                                          Feet.

2 sills,	6 x 6 inches,	16 feet long,		96
2 sills,	6 x 6 "	10 "		60
12 first floor joists,	2 x 6 inches.	10 feet long,		120
13 second "	" 2 x 6 "	10 "		130
4 posts,	4 x 4 "	12 "		64
48 studs,	2 x 4 "	12 "		384
18 rafters,	2 x 5 "	7 "		105
		Total number of feet,	———	959

Enclosing and under floor boards, 1236 feet.
               Top floor boards, 320 feet.
Shingles, 2 M. Clapboards, 830 feet.

BALLOON FRAMES.

No. 13 building. 12 by 12 feet, posts 8 feet in height.

				Feet.
4 sills,	6 x 6 inches,	12 feet long,	144
9 floor joists,	2 x 6 "	12 "	108
4 posts,	4 x 4 "	8 "	43
46 studs,	2 x 4 "	8 "	245
4 tie studs and plates,	2 x 4 "	12 "	32
14 rafters,	2 x 5 "	8 "	94

Total number of feet, ——— 666

Enclosing and under floor boards, 802 feet.
Top floor boards, 144 feet.
Shingles, 1¾ M. Clapboards, 540 feet.

No. 14 building, 12 by 12 feet, posts 10 feet in height.
Same as No. 13, with the additional length of posts and studding.

Total number of feet, 736

Enclosing and under floor boards, 900 feet.
Top floor boards, 144 feet.
Shingles, 1¾ M. Clapboards, 660 feet.

No. 15 building, 12 by 12 feet, posts 12 feet in height.

				Feet.
4 sills,	6 x 6 inches,	12 feet long,	144
9 first floor joists,	2 x 6 "	12 "	108
10 second floor "	2 x 6 "	12 "	120
4 posts,	4 x 4 "	12 "	64
43 studs,	2 x 4 "	12 "	344
14 rafters,	2 x 5 "	8 "	94

Total number of feet, ——— 874

Enclosing and under floor boards, 1128 feet.
Top floor boards, 288 feet.
Shingles, 1¾ M. Clapboards, 800 feet.

BALLOON FRAMES.

No. 16 building, 12 by 14 feet, posts 8 feet in height.

			Feet.
2 sills	6 x 6 inches, 14 feet long,		84
2 "	6 x 6 " 12 "		72
10 floor joists;	2 x 6 " 12 "		120
4 posts,	4 x 4 " 8 "		43
50 studs,	2 x 4 " 8 "		267
4 tie studs and plates,	2 x 4 " 14 "		37
16 rafters,	2 x 5 " 8 "		106
	Total number of feet,		——— 729

Enclosing and under floor boards, 925 feet.
Top floor boards, 168 feet.
Shingles, 2 M.   Clapboards, 580 feet.

No. 17 building, 12 by 14 feet, posts 10 feet in height.
Same as No. 16, with the additional length of posts and studding.

Total number of feet,   800

Enclosing and under floor boards, 992 feet.
Top floor boards, 168 feet.
Shingles, 2 M.   Clapboards, 710 feet.

No. 18 building, 12 by 14 feet, posts 12 feet in height.

			Feet.
2 sills,	6 x 6 inches, 14 feet long,		84
2 "	6 x 6 " 12 "		72
10 1st. floor joists,	2 x 6 " 12 "		120
11 2d. "	2 x 6 " 12 "		132
4 posts,	4 x 4 " 12 "		64
46 studs,	2 x 4 " 12 "		368
16 rafters.	2 x 5 " 8 "		106
	Total number of feet,		——— 946

Enclosing and under floor boards, 1264 feet.
Top floor boards, 336 "
Shingles, 2 M.   Clapboards, 840 feet.

BALLOON FRAMES.

No. 19 building, 12 by 16 feet, posts 8 feet in height.

				Feet.
2 sills,	6 x 6 inches,	16 feet long,	96
2 sills,	6 x 6 "	12 "	72
12 floor joists,	2 x 6 "	12 "	144
4 posts,	4 x 4 "	8 "	43
54 studs,	2 x 4 "	8 "	288
4 tie studs and plates,	2 x 4 "	16 "	43
18 rafters,	2 x 5 "	8 "	120

Total number of feet, ——— 806

Enclosing and under floor boards, 978 feet.
Top floor boards, 192 feet.
Shingles, about 2¼ M. Clapboards, 620 feet.

No. 20 building, 12 by 16 feet, posts 10 feet in height.
Same as No. 19, with the additional length of posts and studding.

Total number of feet, 900

Enclosing and under floor boards, 1090 feet.
Top floor boards, 192 feet.
Shingles, about 2¼ M. Clapboards, 720 feet.

No. 21 building. 12 by 16 feet, 12 feet in height.

				Feet.
2 sills,	6 x 6 inches,	16 feet long,	96
5 sills,	6 x 6 "	12 "	72
12 1st floor joists	2 x 6 "	12 "	144
13 2d floor joists	2 x 6 "	12 "	156
4 posts,	4 x 4 "	12 "	64
52 studs,	2 x 4 "	12 "	416
18 rafters,	2 x 5 "	8 "	120

Total number of feet, ——— 1068

Enclosing and under floor boards, 1394 feet.
Top floor boards, 384 feet.
Shingles, about 2¼ M. Clapboards, 900 feet.

BALLOON FRAMES.

No. 22 building, 14 by 14 feet, posts 8 feet in height.

		Feet.
4 sills,	7 x 7 inches, 14 feet long,	228
10 floor joists,	2 x 7 " 14 "	163
4 posts,	4 x 6 " 8 "	64
56 studs,	2 x 4 " 8 "	298
4 tie studs and plates	2 x 4 " 14 "	37
16 rafters,	2 x 5 " 9½ "	128
	Total number of feet	——— 918

Enclosing and under floor boards, 1018 feet.
Top floor boards, 196 feet.
Shingles, 2⅓ M. Clapboards, 647 feet.

No. 23 building, 14 by 14 feet, posts 12 feet in height.

		Feet.
4 sills,	7 x 7 inches, 14 feet long,	228
10 1st floor joists	2 x 7 " 14 "	163
11 2d floor joists,	2 x 6 " 14 "	154
4 posts.	4 x 6 " 12 "	96
53 studs,	2 x 4 " 12 "	424
16 rafters,	2 x 5 " 9½ "	128
	Total number of feet,	——— 1193

Enclosing and under floor boards, 1438 feet.
Top floor boards, 392 feet.
Shingles, 2⅓ M. Clapboards, 927 feet,

No. 24 building, 14 by 14 feet, posts 14 feet in height.
Same as No. 23 with the additional length of posts and studding.

Total number of feet, 1300

Enclosing and under floor boards, 1550 feet.
Top floor boards, 392 feet.
Shingles, 2⅓ M. Clapboards, 1067 feet.

BALLOON FRAMES.

No. 25 building, 14 by 16 feet, posts 8 feet in height.

				Feet.
2 sills,	7 x 7 inches,	16 feet long,	131
2 sills,	7 x 7 "	14 "	114
12 floor joists,	2 x 7 "	14 "	196
4 posts,	4 x 6 "	8 "	64
60 studs,	2 x 4 "	8 "	320
4 tie studs and plates,	2 x 4 "	16 "	43
18 rafters,	2 x 5 "	9½ "	144

Total number of feet, ——— 1012

Enclosing and under floor boards, 1121 feet.
Top floor boards, 224 feet.
Shingles, 2⅝ M. Clapboards, 687 feet.

No. 26 building, 14 by 16 feet, posts 12 feet in height.

				Feet.
2 sills,	7 x 7 inches,	16 feet long,	131
2 sills,	7 x 7 "	14 "	114
4 posts,	4 x 6 "	12 "	96
12 first floor joists,	2 x 7 "	14 "	196
13 second floor "	2 x 6 "	14 "	162
53 studs,	2 x 4 "	12 "	424
4 studs,	2 x 4 "	16 "	43
18 rafters,	2 x 5 "	9½ "	144

Total number of feet, ——— 1310

Enclosing and under floor boards, 1585 feet.
Top floor boards, 448 feet.
Shingles, 2⅝ M. Clapboards, 993 feet.

No. 27 building, 14 by 16 feet, posts 14 feet in height.

Same as No. 26, with the additional length of posts and studding.

Total number of feet, 1390

Enclosing and under floor boards, 1705 feet.
Top floor boards, 448 feet.
Shingles, 2⅝ M. Clapboards, 1143 feet.

BALLOON FRAMES.

No. 28 building 14 by 18 feet, posts 8 feet in height.

				Feet.
2 sills,	7 x 7 inches,	18 feet long,		147
2 "	7 " 7 "	14 " "		114
13 floor joists,	2 " 7 "	14 " "		212
4 posts,	4 " 6 "	8 " "		64
63 studs,	2 " 4 "	8 " "		336
5 tie studs & plates,	2 " 4 "	14 " "		46
20 rafters,	2 " 5 "	9½ " "		160

Total number of feet, ——— 1079

Enclosing and under floor boards, 1219 feet.
Top floor boards, 252 "
Shingles, 3 M. Clapboards, 733 feet.

No. 29 building, 14 by 18 feet, posts 12 feet in height

				feet.
2 sills,	7 x 7 inches,	18 feet long,		147
2 "	7 x 7 "	14 "		114
13 1st floor joists	2 x 7 "	14 "		212
14 2d. "	2 x 6 "	14 "		196
4 posts,	4 x 6 "	12 "		96
57 studs,	2 x 4 "	12 "		456
5 tie studs & plates,	2 x 4 "	14 "		46
20 rafters,	2 x 5 "	9½ "		160

Total number of feet, ——— 1427

Enclosing and under floor boards, 1727 feet.
Top floor boards, 504 "
Shingles 3 M. Clapboards, 1054 feet.

No. 30 building, 14 by 18 feet, posts 14 feet in height.

Same as No. 29, with the additional length of posts and studding

Total number of feet, .1514.

Enclosing and under floor boards, 1855 feet.
Top floor boards, 504 "
Shingles 3 M. Clapboards, 1213 feet.

BALLOON FRAMES.

No. 31 building, 14 by 20 feet, posts 8 feet in height.

						feet.
2 sills,	7 x 7 inches,	20 feet long,				163
2 sills,	7 x 7	"	14	"		114
15 floor joists,	2 x 7	"	14	"		245
4 posts,	4 x 6	"	8	"		64
66 studs,	2 x 4	"	8	"		352
5 tie studs & plates,	2 x 4	"	14	"		46
22 rafters,	2 x 5	"	9½	"		176

Total number of feet, —— 1160

Enclosing and under floor boards, 1317 feet
Top floor boards, 280 feet.
Shingles, 3¼ M. Clapboards, 773 feet.

No. 32 building, 14 by 20 feet, posts 12 feet in height.

						feet.
2 sills,	7 x 7 inches,	20 feet long,				163
2 sills,	7 x 7	"	14	"		114
15 1st floor joists	2 x 7	"	14	"		245
16 2d. "	2 x 6	"	14	"		224
4 posts,	4 x 6	"	12	"		96
60 studs,	2 x 4	"	12	"		480
5 tie studs & plates,	2 x 4	"	14	"		46
22 rafters,	2 x 5	"	9½	"		176

Total number of feet, ——– 1544

Enclosing and under floor boards, 1869 feet.
Top floor boards 560 feet.
Shingles, 3¼ M. Clapboards, 1107 feet.

No. 33 building, 14 by 20 feet, posts 14 feet in height.
Same as No. 32, with the additional length of posts and studding.

Total number of feet, 1640

Enclosing and under floor boards, 2005 feet.
Top floor boards, 560 feet.
Shingles, 3¼ M. Clapboards, 1114 feet.

BALLOON FRAMES.

No. 34 building, 16 by 16 feet, posts 8 feet in height.

			Feet.
4 sills,	7 x 7 inches,	16 feet long,	261
12 floor joists,	2 x 7 "	16 "	228
4 posts,	4 x 6 "	8 "	64
63 studs,	2 x 4 "	8 "	336
5 tie studs & plates,	2 x 4 "	16 "	61
18 rafters,	2 x 6 "	11 "	198

Total number of feet, ——— 1148

Enclosing and under floor boards, 1260 feet,
Top floor boards, 256 feet.
Shingles, 3 M. Clapboards, 760 feet.

No. 35 building, 16 by 16 feet, posts 12 feet in height.

			Feet.
4 sills,	7 x 7 inches,	16 feet long,	261
12 first floor joists,	2 x 7 inches,	16 feet long,	228
13 second "	" 2 x 6 "	16 "	208
4 posts,	4 x 6 "	12 "	96
60 studs,	2 x 4 "	12 "	480
5 tie studs & plates,	2 x 4 "	16 "	61
18 rafters,	2 x 6 "	11 "	198

Total number of feet, ——— 1532

Enclosing and under floor boards, 1772 feet.
Top floor boards, 512 feet.
Shingles, 3 M. Clapboards, 1082 feet.

No. 36 building, 16 by 16 feet, posts 14 feet in height.
Same as No. 35, with the additional length of posts and studding.

Total number of feet, 1626

Enclosing and under floor boards, 1900 feet.
Top floor boards, 512 feet.
Shingles, 3 M. Clapboards, 1240 feet.

BALLOON FRAMES.

No. 37 building, 16 by 18 feet, posts 8 feet in height.

			feet.
2 sills,	7 x 7 inches,	18 feet long,	147
2 "	7 x 7 "	16 "	130
13 floor joists,	2 x 7 "	16 "	242
4 posts,	4 x 6 "	8 "	64
65 studs.	2 x 4 "	8 "	347
3 tie studs	2 x 4 "	16 "	32
2 plates	2 x 6 "	18 "	36
20 rafters,	2 x 6 "	11 "	220
	Total number of feet,		—— 1218

Enclosing and under floor boards, 1400 feet.
Top floor boards, 288 feet.
Shingles, 3½ M. Clapboards, 800 feet.

No. 38 building, 16 by 18 feet, posts 14 feet in height.

			feet.
2 sills,	7 x 7 inches,	18 feet long,	147
2 "	7 x 7 "	16 "	130
13 1st floor joists	2 x 7 "	16 "	242
14 2d. "	2 x 6 "	16 "	224
4 posts,	4 x 6 "	14 "	112
60 studs,	2 x 4 "	14 "	560
3 tie studs	2 x 4 "	16 "	32
2 plates	2 x 6 "	18 "	36
20 rafters,	2 x 6 "	11 "	220
	Total number of feet,		—— 1703

Enclosing and under floor boards, 2128 feet.
Top floor boards, 576 feet.
Shingles, about 3½ M. Clapboards, 1310 feet.

No. 38. building, 16 by 18 feet, posts 20 feet in height.
Same as No. 38 with the additional length of posts and studding and 14 upper floor joists.

Total number of feet, 2215
Enclosing and under floor boards, 2824 feet.
Top floor boards, 576 feet.
Shingles, about 3½ M. Clapboards, 1820 feet.

BALLOON FRAMES.

No. 40 building, 16 by 20 feet, posts 8 feet in height.

				Feet.
3 sills	7 x 7 inches,	16 feet long,	195
4 "	7 x 7 "	10 "	163
14 floor joists;	2 x 7 "	16 "	264
6 posts,	4 x 6 "	8 "	96
68 studs,	2 x 4 "	8 "	363
3 tie studs,	2 x 4 "	16 "	32
2 plates,	2 x 6 "	20 "	40
22 rafters,	2 x 6 "	11 "	242
	Total number of feet,		—— 1395

Enclosing and under floor boards, 1476 feet.
Top floor boards, 320 feet.
Shingles, 3¾ M. Clapboards, 840 feet.

No. 41 building, 16 by 20 feet, posts 14 feet in height.

			Feet.
3 sills,	7 x 7 inches,	16 feet long,	195
4 "	7 x 7 "	10 "	163
14 1st. floor joists,	2 x 7 "	16 "	264
17 2d. "	2 x 6 "	16 "	272
6 posts,	4 x 6 "	14 "	168
65 studs,	2 x 4 "	14 "	607
2 plates,	2 x 6 "	20 "	40
22 rafters	2 x 6 "	11 "	242
	Total number of feet,		—— 1951

Enclosing and under floor boards, 2228 feet.
Top floor boards, 640 "
Shingles, 3¾ M. Clapboards, 1380 feet.

No. 42 building, 16 by 20 feet, posts 20 feet in height.

Same as No. 41, with the additional length of posts, studding and 17 upper floor joists.

Total number of feet, 2488

Enclosing and under floor boards. 2980 feet.
Top floor boards. 640 feet.
Shingles, 3¾ M. Clapboards, 1920 feet.

BALLOON FRAMES.

No. 43 building, 16 by 22 feet, posts 8 feet in height.

				feet
3 sills,	7 x 7 inches,	16 feet long,		195
4 "	7 x 7 "	11 "		180
15 floor joists,	2 x 7 "	16 "		280
6 posts,	4 x 6 "	8 "		96
72 studs.	2 x 4 "	8 "		384
3 tie studs	2 x 4 "	16 "		32
2 plates	4 x 6 "	22 "		88
24 rafters,	2 x 6 "	11 "		264

Total number of feet, —— 1519

Enclosing and under floor boards, 1606 feet.
Top floor boards, 352 feet.
Shingles, 4¼ M. Clapboards, 880 feet.

No. 44 building, 16 by 22 feet, posts 14 feet in height.

				feet
3 sills,	7 x 7 inches,	16 feet long,		195
4 "	7 x 7 "	11 "		180
15 1st floor joists	2 x 7 "	16 "		280
18 2d. "	2 x 7 "	16 "		336
6 posts,	4 x 6 "	14 "		168
67 studs,	2 x 4 "	14 "		625
24 rafters,	2 x 6 "	11 "		264
2 plates	4 x 6 "	22 "		88

Total number of feet, —— 2136

Enclosing and under floor boards, 2414 feet.
Top floor boards, 704 feet.
Shingles, about 4¼ M. Clapboards, 1450 feet.

No 45. building, 16 by 22 feet, posts 20 feet in height.

Same as No. 44 with the additional length of posts, studding and 18 upper floor joists.

Total number of feet, 2744

Enclosing and under floor boards, 3228 feet.
Top floor boards, 704 feet.
Shingles, about 4¼ M. Clapboards, 2000 feet.

BALLOON FRAMES.

No. 46 building. 16 by 24 feet, posts 8 feet in height.

			Feet.
3 sills,	7 x 7 inches,	16 feet long,	195
4 sills,	7 x 7 "	12 "	196
16 floor joists,	2 x 7 "	16 "	298
6 posts,	4 x 6 "	8 "	96
75 studs,	2 x 4 "	8 "	400
3 tie studs,	2 x 4 "	16 "	32
2 plates,	4 x 6 "	24 "	96
26 rafters,	2 x 6 "	11 "	286

Total number of feet, ——— 1599

Enclosing and under floor boards, 1714 feet.
Top floor boards, 384 feet.
Shingles, 4½ M. Clapboards, 920 feet.

No. 47 building, 16 by 24 feet, posts 14 feet in height.

			Feet.
3 sills,	7 x 7 inches,	16 feet long,	195
4 sills,	7 x 7 "	12 "	196
16 first floor joists,	2 x 7 "	16 "	298
19 second floor "	2 x 7 "	18 "	354
6 posts,	4 x 6 "	14 "	168
70 studs,	2 x 4 "	.14 "	653
2 plates,	4 x 6 "	24 "	96
26 rafters,	2 x 6 "	11 "	286

Total number of feet, ——— 2246

Enclosing and under floor boards, 2578 feet.
Top floor boards, 768 feet.
Shingles, 4½ M. Clapboards, 1520 feet.

No. 48 building, 16 by 24 feet, posts 20 feet in height.

Same as No. 47 with the additional length of posts, studding and 19 upper floor joists.

Total number of feet,   2885

Enclosing and under floor boards. 3442 feet.
Top floor boards, 768 feet.
Shingles, 4½ M. Clapboards, 2120 feet.

BALLOON FRAMES.

No. 49 building 18 by 18 feet, posts 8 feet in height.

Feet.
3 sills,	7 x 7 inches,	18 feet long,	220
4 "	7 " 7 "	9 " "	147
26 floor joists,	2 " 7 "	9 " "	273
6 posts,	4 " 6 "	8 " "	96
75 studs,	2 " 4 "	8 " "	400
3 tie studs,	2 " 6 "	18 " "	54
2 plates,	4 x 6 "	18 " "	72
20 rafters,	2 " 6 "	12½ " "	250

Total number of feet, —— 1512

Enclosing and under floor boards, 1542 feet.

Top floor boards, 324 "

Shingles, 4 M. Clapboards, 865 feet.

No. 50 building, 18 by 18 feet, posts 14 feet in height

feet.
3 sills,	7 x 7 inches,	18 feet long,	220
4 "	7 x 7 "	9 "	147
26 1st floor joists	2 x 7 "	9 "	273
15 2d. "	2 x 7 "	18 "	315
6 posts,	4 x 6 "	14 "	168
73 studs,	2 x 4 "	14 "	681
2 plates,	4 x 6 "	18 "	72
20 rafters,	2 x 6 "	12½ "	250

Total number of feet, —— 2126

Enclosing and under floor boards, 2298 feet.

Top floor boards, 648 "

Shingles 4 M. Clapboards, 1400 feet.

No. 51 building, 18 by 18 feet, posts 20 feet in height.

Same as No. 50, with the additional length of posts, studding and 15 upper floor joists.

Total number of feet, 2765.

Enclosing and under floor boards, 3054 feet.

Top floor boards, 648 "

Shingles 4 M. Clapboards, 1945 feet.

BALLOON FRAMES.

No. 52 building, 18 by 20 feet, posts 8 feet in height.

			feet.
3 sills,	7 x 7 inches,	18 feet long,	220
4 sills,	7 x 7 "	10 "	163
26 floor joists,	2 x 7 "	10 "	302
6 posts,	4 x 6 "	8 ".	96
78 studs,	2 x 4 "	8 "	416
3 tie studs	2 x 6 "	18 "	54
2 plates,	4 x 6 "	20 "	80
22 rafters,	2 x 6 "	12½ "	275

Total number of feet, —— 1606

Enclosing and under floor boards, 1668 feet

Top floor boards, 360 feet.

Shingles, 4½ M. Clapboards, 917 feet.

No. 53 building, 18 by 20 feet, posts 14 feet in height.

			feet.
3 sills,	7 x 7 inches,	18 feet long,	220
4 sills,	7 x 7 "	10 "	163
26 1st floor joists	2 x 7 "	10 "	302
16 2d. "	2 x 7 "	18 "	336
6 posts,	4 x 6 "	14 "	168
75 studs,	2 x 4 "	14 "	700
2 plates,	4 x 6 "	20 "	80
22 rafters,	2 x 6 "	12½ "	275

Total number of feet, ——- 2244

Enclosing and under floor boards, 2484 feet.

Top floor boards 720 feet.

Shingles, 4½ M. Clapboards, 1487 feet.

No. 54 building, 18 by 20 feet, posts 20 feet in height.

Same as No. 53, with the additional length of posts, studding and 16 upper floor joists.

Total number of feet, 2925

Enclosing and under floor boards, 3300 feet.

Top floor boards, 720 feet.

Shingles, 4½ M. Clapboards, 2057 feet.

BALLOON FRAMES.

No. 55 building, 18 by 22 feet, posts 8 feet in height.

			Feet.
3 sills,	7 x 7 inches, 18 feet long,	220
4 sills,	7 x 7 " 11 "	180
26 floor joists,	2 x 7 " 11 "	330
6 posts,	4 x 7 " 8 "	96
82 studs,	2 x 4 " 8 "	437
3 tie studs	2 x 6 " 18 "	54
2 plates	4 x 6 " 22 "	88
24 rafters,	2 x 5 " 12½ "	300
Total number of feet		——— 1705

Enclosing and under floor boards, 1786 feet.
Top floor boards, 396 feet.
Shingles, 4⅞ M. Clapboards, 957 feet.

No. 56 building, 18 by 22 feet, posts 14 feet in height.

			Feet.
3 sills,	7 x 7 inches, 18 feet long,	220
4 sills,	7 x 7 " 11 "	180
26 1st floor joists	2 x 7 " 11 "	330
18 2d floor joists,	2 x 7 " 18 "	378
6 posts.	4 x 6 " 14 "	168
78 studs,	2 x 4 " 14 "	728
2 plates	4 x 6 " 22 "	88
24 rafters,	2 x 5 " 12½ "	300
Total number of feet,		——— 2392

Enclosing and under floor boards, 2662 feet.
Top floor boards, 792 feet.
Shingles. 4⅞ M. Clapboards, 1557 feet,

No. 57 building, 18 by 22 feet, posts 20 feet in height.
Same as No. 56 with the additional length of posts and studding and 18 upper floor joists.

Total number of feet, 3100

Enclosing and under floor boards, 3536 feet.
Top floor boards, 792 feet.
Shingles, 4⅞ M. Clapboards, 2157 feet.

BALLOON FRAMES.

No. 58 building, 18 by 24 feet, posts 8 feet in height.

				Feet.	
3 sills,	7 x 7 inches,	18 feet long,	220		
4 sills,	7 x 7 "	12 "	196		
26 floor joists,	2 x 7 "	12 "	364		
6 posts,	4 x 6 "	8 "	96		
85 studs,	2 x 4 "	8 "	453		
4 tie studs	2 x 6 "	18 "	72		
2 plates,	4 x 6 "	24 "	96		
26 rafters,	2 x 6 "	12½ "	325		

Total number of feet, ——— 1822

Enclosing and under floor boards, 1904 feet,
Top floor boards, 412 feet.
Shingles, 5¼ M. Clapboards, 997 feet.

No. 59 building, 18 by 24 feet, posts 14 feet in height.

				Feet.
3 sills,	7 x 7 inches,	18 feet long,	220	
4 sills,	7 x 7 "	12 "	196	
26 first floor joists,	2 x 7 inches,	12 feet long,	364	
19 second "	" 2 x 7 "	18 "	399	
6 posts,	4 x 6 "	14 "	168	
82 studs,	2 x 4 "	14 "	765	
2 plates,	4 x 6 "	24 "	96	
26 rafters,	2 x 6 "	12½ "	325	

Total number of feet, ——— 2533

Enclosing and under floor boards, 2840 feet.
Top floor boards, 824 feet.
Shingles, 5¼ M. Clapboards, 1637 feet.

No. 60 building, 18 by 24 feet, posts 20 feet in height.

Same as No. 59, with the additional length of posts, studding and 19 upper floor joists.

Total number of feet, 3270

Enclosing and under floor boards, 3776 feet.
Top floor boards, 824 feet.
Shingles, 5¼ M. Clapboards, 2257 feet.

BALLOON FRAMES.

No. 61 building, 18 by 26 feet, posts 8 feet in height.

				Feet.
3 sills,	7 x 7 inches,	18 feet long,	220
4 sills,	7 x 7 "	13 "	212
26 floor joists,	2 x 7 "	13 "	394
6 posts,	4 x 6 "	8 "	96
88 studs,	2 x 4 "	8 "	469
4 tie studs	2 x 6 "	18 "	72
2 plates,	4 x 6 "	26 "	104
28 rafters,	2 x 6 "	12½ "	350

Total number of feet, ——— 1913

Enclosing and under floor boards, 2022 feet.
Top floor boards, 468 feet.
Shingles, 5½ M. Clapboards, 1037 feet.

No. 62 building. 18 by 26 feet, 14 feet in height.

				Feet.
3 sills,	7 x 7 inches,	18 feet long,	220
4 sills,	7 x 7 "	13 "	212
26 1st floor joists	2 x 7 "	13 "	394
21 2d floor joists	2 x 7 "	18 "	441
84 studs,	2 x 4 "	14 "	784
2 plates,	4 x 6 "	26 "	104
28 rafters,	2 x 6 "	12½ "	350
6 posts,	4 x 6 "	14 "	168

Total number of feet, ——— 2673

Enclosing and under floor boards, 3018 feet.
Top floor boards, 936 feet.
Shingles, 5½ M. Clapboards, 1697 feet.

No. 63 building, 18 by 26 feet, posts 20 feet in height.

Same as No. 62, with the additional length of posts, studding and 21 upper floor joists.

Total number of feet, 3436

Enclosing and under floor boards, 4014 feet.
Top floor boards, 936 feet.
Shingles, 5½ M. Clapboards, 2357 feet.

BALLOON FRAMES.

No. 64 building, 18 by 28 feet, posts 8 feet in height.

			Feet.
3 sills	7 x 7 inches,	18 feet long,	220
4 "	7 x 7 "	14 "	237
26 floor joists;	2 x 7 "	14 "	424
6 posts,	4 x 6 "	8 "	96
92 studs,	2 x 4 "	8 "	490
4 tie studs,	2 x 6 "	18 "	72
2 plates,	4 x 6 "	28 "	112
30 rafters,	2 x 6 "	12½ "	375

Total number of feet, ——— 2026

Enclosing and under floor boards, 2140 feet.
Top floor boards, 504 feet.
Shingles, 5⅞ M. Clapboards, 1077 feet.

No. 65 building, 18 by 28 feet, posts 14 feet in height.

			Feet.
3 sills,	7 x 7 inches,	18 feet long,	220
4 "	7 x 7 "	14 "	237
26 1st. floor joists,	2 x 7 "	14 "	424
22 2d. "	2 x 7 "	18 "	462
6 posts,	4 x 6 "	14 "	168
87 studs,	2 x 4 "	14 "	812
2 plates,	4 x 6 "	28 "	112·
30 rafters	2 x 6 "	12½ "	375

Total number of feet, ——— 2810

Enclosing and under floor boards, 3196 feet.
Top floor boards, 1008 "
Shingles, 5⅞ M. Clapboards, 1767 feet.

No. 66 building, 18 by 28 feet, posts 20 feet in height.

Same as No. 65, with the additional length of posts, studding and 22 upper floor joists.

Total number of feet, 3645

Enclosing and under floor boards, 4252 feet.
Top floor boards, 1008 feet.
Shingles, 5⅞ M. Clapboards, 2457 feet.

5

BALLOON FRAMES.

No. 67 building 18 by 30 feet, posts 8 feet in height.

				Feet.
3 sills,	7 x 7 inches,	18 feet long,		220
4 "	7 " 7 "	15 " "		245
26 floor joists,	2 " 7 "	15 " "		451
6 posts,	4 " 6 "	8 " "		96
95 studs,	2 " 4 "	18 " "		507
4 tie studs,	2 " 6 "	18 " "		72
2 plates,	4 x 6 "	30 " "		120
32 rafters,	2 " 6 "	12½ " "		400

Total number of feet, —— 2111

Enclosing and under floor boards, 2258 feet.

Top floor boards, 540 "

Shingles, 6⅓ M. Clapboards, 1117 feet.

No. 68 building, 18 by 30 feet, posts 14 feet in height

				feet..
3 sills,	7 x 7 inches,	18 feet long,		220
4 "	7 x 7 "	15 "		245
26 1st floor joists	2 x 7 "	15 "		451
24 2d. "	2 x 7 "	18 "		504
6 posts,	4 x 6 "	14 "		168
90 studs,	2 x 4 "	14 "		840
2 plates,	4 x 6 "	30 "		120
32 rafters,	2 x 6 "	12½ "		400

Total number of feet, —— 2948

Enclosing and under floor boards, 3374 feet.

Top floor boards, 1080 "

Shingles 6⅓ M. Clapboards, 1832 feet.

No. 69 building, 18 by 30 feet, posts 20 feet in height.

Same as No. 68, with the additional length of posts, studding and 24 upper floor joists.

Total number of feet, 2825.

Enclosing and under floor boards, 4490 feet.

Top floor boards, 1080 "

Shingles 6⅓ M. Clapboards, 2557 feet.

BALLOON FRAMES.

No. 70 building, 20 by 20 feet, posts 12 feet in height.

					feet.
3 sills,	7 x 7 inches,	20 feet long,	245
4 sills,	7 x 7 "	10 "	163
30 floor joists,	2 x 7 "	10 "	350
16 2nd "	2 x 7 "	20 "	373
6 posts,	4 x 6 "	12 "	144
87 studs,	2 x 4 "	12 "	696
2 plates,	4 x 6 "	20 "	80
22 rafters,	2 x 6 "	14 . "	308

Total number of feet, —— 2359
Enclosing and under floor boards, 2564 feet
Top floor boards, 800 feet.
Shingles, 5 M. Clapboards, 1400 feet.

No. 71 building, 20 by 20 feet, posts 14 feet in height.
Same as No. 70, posts and studding 2 feet longer.
Total number of feet, 2500
Enclosing and under floor boards, 2724 feet.
Top floor boards, 800 feet.
Shingles, 5 M. Clapboards, 1600 feet.

No. 72 building, 20 by 20 feet, posts 20 feet in height.

					feet.
3 sills,	7 x 7 inches,	20 feet long,	245
4 sills,	7 x 7 "	10 "	163
30 1st floor joists	2 x 7 "	10 "	350
16 2d. "	2 x 7 "	20 "	373
16 3d. "	2 x 7 "	20 "	373
6 posts,	4 x 6 "	20 "	240
60 studs,	2 x 4 "	20 "	800
23 gable & extra studs,	2 x 4 "	12 "	184
2 plates,	4 x 6 "	20 "	80
22 rafters,	2 x 6 "	14 "	308

Total number of feet, —— 3116
Enclosing and under floor boards, 3604 feet.
Top floor boards 1200 feet.
Shingles, 5 M. Clapboards, 2200 feet.

BALLOON FRAMES.

No. 73 building, 20 by 22 feet, posts 12 feet in height.

			Feet.
3 sills,	7 x 7 inches,	20 feet long,	245
4 sills,	7 x 7 "	11 "	180
30 1st floor joists,	2 x 7 "	11 "	382
17 2nd " "	2 x 7 "	20 "	396
6 posts,	4 x 6 "	12 "	144
90 studs,	2 x 4 "	12 "	720
2 plates,	4 x 6 "	22 "	88
24 rafters,	2 x 6 "	14 "	336

Total number of feet, ——— 2491

Enclosing and under floor boards, 2748 feet.
Top floor boards, 880 feet.
Shingles, 5½ M. Clapboards, 1460 feet.

No. 74 building, 20 by 22 feet, posts 14 feet in height.

Same as No. 73, posts and studding 2 feet longer.
Total number of feet,   2635
Enclosing and under floor boards, 2916 feet.
Top floor boards, 880 feet.
Shingles, 5½ M. Clapboards, 1667 feet.

No. 75 building. 20 by 22 feet, posts 20 feet in height.

			Feet.
3 sills,	7 x 7 inches,	20 feet long,	245
4 sills,	7 x 7 "	11 "	180
30 1st floor joists	2 x 7 "	11 "	382
17 2d floor joists	2 x 7 "	20 "	396
18 3d " "	2 x 6 "	20 "	360
6 posts,	4 x 6 "	20 "	240
64 studs,	2 x 4 "	20 "	832
25 gable & extra studs,	2 x 4 "	12 "	200
2 plates,	4 x 6 "	22 "	88
24 rafters,	2 x 6 "	14 "	336

Total number of feet, ———3259

Enclosing and under floor boards, 3892 feet.
Top floor boards, 1320 feet.
Shingles, 5½ M. Clapboards, 2340 feet.

BALLOON FRAMES.

No. 76 building, 20 by 24 feet, posts 12 feet in height.

				feet.	
3 sills,	7 x 7 inches,	20 feet long,	245	
4 "	7 x 7 "	12 "	196	
30 1st floor joists,	2 x 7 "	12 "	420	
19 2d floor joists,	2 x 7 "	20 "	437	
6 posts,	4 x 6 "	12 "	144	
2 plates	4 x 6 "	24 "	96	
26 rafters,	2 x 6 "	14 "	364	
97 studs.	2 x 4 "	12 "	776	
collar girts,	2 x 6 "	about	100	
		Total number of feet,	—— 2778	

Enclosing and under floor boards, 2932 feet.
Top floor boards, 960 feet.
Shingles, 5⅞ M.   Clapboards, 1510 feet.

———

No. 77. building, 20 by 24 feet, posts 14 feet in height.

Same as No. 76 with the additional length of posts, and studding.

Total number of feet, 2931

Enclosing and under floor boards, 3108 feet.
Top floor boards, 960 feet.
Shingles, about 5⅞ M.   Clapboards, 1740 feet.

BALLOON FRAMES.

No. 78 building, 20 by 24 feet, posts 20 feet in height

						Feet.
3 sills,	7 x 7 inches,	20 feet long,	245
4 "	7 x 7 "	12 "	196
30 1st floor joists	2 x 7 "	12 "	420
19 2d. "	2 x 7 "	20 "	437
19 3d. "	2 x 6 "	20 "	380
6 posts,	4 x 6 "	20 "	242
2 plates	4 x 6 "	24 "	96
26 rafters,	2 x 6 "	14 "	364
66 studs,	2 x 4 "	20 "	880
26 gable & extra studs	2 x 4 "	12 "	208
13 collar girts,	2 x 6 "		100

Total number of feet, —— 3568

Enclosing and under floor boards, 4096 feet.
Top floor boards, 1440 feet.
Shingles, about 5⅞ M. Clapboards, 2400 feet.

No. 79 building, 20 by 26 feet, posts 12 feet in height.

						Feet.
3 sills,	7 x 7 inches,	20 feet long,	245
4 sills,	7 x 7 "	13 "	212
30 first floor joists,	2 x 7 "	13 "	455
20 second floor "	2 x 7 "	20 "	466
6 posts,	4 x 6 "	12 "	144
100 studs,	2 x 4 "	12 "	800
28 rafters,	2 x 6 "	14 "	392
plates,	4 x 6 "	52 "	104
collar girts,	2 x 6 "		100

Total number of feet, —— 2918

Enclosing and under floor boards, 3116 feet.
Top floor boards, 1040 feet.
Shingles, 6¼ M. Clapboards, 1580 feet.

BALLOON FRAMES.

No. 80 building, 20 by 26 feet, posts 14 feet in height.

Same as No. 79, posts and studding 2 feet longer.
  Total number of feet, 3075
Enclosing and under floor boards, 3300 feet.
  Top floor boards, 1040 feet.
Shingles, 6¼ M. Clapboards, 1810 feet.

No. 81 building. 20 by 26 feet, posts 20 feet in height.

				Feet.
3 sills,	7 x 7 inches,	20 feet long,	245
4 sills,	7 x 7 "	13	"	212
30 first floor joists,	2 x 7 "	13	"	445
20 second floor joists,	2 x 7 "	20	"	466
20 third floor joists,	2 x 6 "	20	"	400
6 posts,	4 x 6 "	20	"	240
70 studs,	2 x 4 "	20	"	933
26 gable and extra studs	2 x 4 "	12	"	208
28 rafters,	2 x 6 "	14	"	392
2 plates,	4 x 6 "	26	"	104
collar girts,	2 x 6 "			100

  Total number of feet, ——— 3745
Enclosing and under floor boards, 4375 feet.
  Top floor boards, 1560 feet.
Shingles, 6¼ M. Clapboards, 2500 feet.

BALLOON FRAMES.

No. 82 building, 20 by 28 feet, posts 12 feet in height.

					Feet.
3 sills,	7 x 7 inches,	20 feet long,	245
4 sills,	7 x 7 "	14 "	229
2 tie sills,	4 x 7 "	14 "	65
28 1st floor joists,	2 x 7 "	14 "	457
22 2d floor joists,	2 x 7 "	20 "	513
6 posts,	4 x 6 "	12 "	144
2 plates,	4 x 6 "	56 "	112
103 studs,	2 x 4 "	12 "	824
30 rafters,	2 x 6 "	14 "	420
collar girts,	2 x 6 "	"	125

Total number of feet, ——— 3134

Enclosing and under floor boards, 3300 feet,
Top floor boards, 1120 feet.
Shingles, 6⅝ M. Clapboards, 1640 feet.

No. 83 building, 20 by 28 feet, posts 14 feet in height.
Same as No. 82, posts and studding 2 feet longer.
Total number of feet, 3292
Enclosing and under floor boards, 3492 feet.
Top floor boards, 1120 feet.
Shingles, 6⅝ M. Clapboards, 1880 feet.

BALLOON FRAMES.

No. 84 building, 20 by 28 feet, posts 20 feet in height.

				Feet.
3 sills,	7 x 7 inches,	20 feet long,		245
4 sills,	7 x 7 "	14 "		229
2 tie sills,	4 x 7 "	14 "		65
28 first floor joists,	2 x 7 inches,	14 feet long,		457
22 second "	" 2 x 7 "	20	"	513
22 third "	" 2 x 6 "	20	"	440
6 posts,	4 x 6 "	20	"	240
2 plates,	4 x 6 "	28	"	112
74 studs,	2 x 4 "	20	"	987
28 gable and extra studs	2 x 4 "	12	"	224
30 rafters,	2 x 6 "	'14	"	420
collar girts,	2 x 6 "			125
		Total number of feet,		——— 4057

Enclosing and under floor boards, 4628 feet.
Top floor boards, 1680 feet.
Shingles, 6⅝ M. Clapboards, 2600 feet.

———

No. 85 building, 20 by 30 feet, posts 12 feet in height.

				Feet.
3 sills,	7 x 7 inches,	20 feet long,		245
6 short sills,	7 x 7 "	15	"	367
28 first floor joists,	2 x 7 "	15	"	490
23 second floor joists,	2 x 7 "	20	"	536
6 posts,	4 x 6 "	12	"	144
plates	4 x 6 "	60	"	120
105 studs,	2 x 4 "	12	"	840
32 rafters,	2 x 6 "	14	"	448
collar girts,	2 x 6			135
		Total number of feet		——— 3325

Enclosing and under floor boards, 3484 feet.
Top floor boards, 1200 feet.
Shingles, 7⅛ M. Clapboards, 1700 feet.

BALLOON FRAMES.

No. 86 building, 20 by 30 feet, posts 14 feet in height.

Same as No. 85, with the additional length of posts and studding.

Total number of feet, 3490

Enclosing and under floor boards, 3684 feet.

Top floor boards, 1200 feet.

Shingles, 7½ M.   Clapboards, 1950 feet.

No. 87 building, 20 by 30 feet, posts 20 feet in height.

					Feet.
3 sills,	7 x 7 inches,	20 feet long,	245
6 sills,	7 x 7 "	15 "	367
28 1st floor joists	2 x 7 "	15 "	490
23 2d floor joists,	2 x 7 "	20 "	536
23 3d floor joists,	2 x 6 "	20 "	460
6 posts.	4 x 6 "	20 "	240
2 plates	4 x 6 "	30 "	120
77 studs	2 x 4 "	20 "	1026
28 gable and extra studs	2 x 4 "	12 "	224
32 rafters,	2 x 6 "	14 "	448
collar girts,	2 x 6 "		135

Total number of feet, ——— 4291

Enclosing and under floor boards, 4884 feet.

Top floor boards, 1800 feet.

Shingles, 7½ M.   Clapboards, 2700 feet,

BALLOON FRAMES.

No. 88 building, 22 by 22 feet, posts 12 feet in height.

			Feet.
3 sills	7 x 7 inches,	22 feet long,	270
6 short sills,	7 x 7 "	11 "	270
32 first floor joists;	2 x 7 "	11 "	408
17 second floor joists,	2 x 7 "	22 "	435
6 posts,	4 x 7 "	12 "	168
plates,	4 x 6 "	44 "	88
97 studs,	2 x 4 "	12 "	776
24 rafters,	2 x 7 "	15 "	420
collar girts,		about	100

Total number of feet, —— 2935

Enclosing and under floor boards, 2954 feet.
Top floor boards, 968 feet.
Shingles, 6 M. Clapboards, 1545 feet.

No. 89 building, 22 by 22 feet, posts 14 feet in height.
Same as No. 88, posts and studding 2 feet longer.

Total number of feet, 3092

Enclosing and under floor boards, 3130 feet.
Top floor boards, 968 feet.
Shingles, 6 M. Clapboards, 1765 feet.

BALLOON FRAMES.

No. 90 building, 22 by 22 feet, posts 20 feet in height.

				Feet.
3 sills,	7 x 7 inches,	22 feet long,	270
6 "	7 x 7 "	11 "	270
32 1st. floor joists,	2 x 7 "	11 "	408
17 2d. "	2 x 7 "	22 "	435
17 3d. floor joists,	2 x 6 "	22 "	374
6 posts,	4 x 7 "	20 "	280
2 plates,	4 x 6 "	22 "	88
67 studs,	2 x 4 "	20 "	893
30 gable and extra studs,	2 x 4 "	12 "	240
24 rafters,	2 x 7 "	15 "	420
collar girts,	2 x 6		about 100
	Total number of feet,		—— 3778

Enclosing and under floor boards, 4142 feet.
Top floor boards, 1452 "
Shingles, 6 M. Clapboards, 2400 feet.

No. 91 building, 22 by 24 feet, posts 12 feet in height.

				feet.
3 sills,	7 x 7 inches,	22 feet long,	270
6 short sills,	7 x 7 "	12 "	294
30 1st floor joists,	2 x 7 "	12 "	420
19 2d floor joists,	2 x 7 "	22 "	486
6 posts,	4 x 7 "	12 "	168
plates,	4 x 6 "	48 "	98
100 studs.	2 x 4 "	12 "	800
26 rafters,	2 x 7 "	15 "	455
collar girts,	2 x 6 "		about 108
	Total number of feet,		—— 3099

Enclosing and under floor boards, 3150 feet.
Top floor boards, 1056 feet.
Shingles, 6¼ M. Clapboards, 1605 feet.

BALLOON FRAMES.

No. 92. building, 22 by 24 feet, posts 14 feet in height.

Same as No. 91 posts and studding 14 feet long.
                          Total number of feet,   3236
Enclosing and under floor boards, 3334 feet.
             Top floor boards, 1056 feet.
Shingles, about 6¼ M. Clapboards, 1835 feet.

No. 93 building, 22 by 24 feet, posts 20 feet in height.

				Feet.
3 sills,	7 x 7 inches,	22	feet long,	270
6 short sills,	7 x 7 "	12	"	294
30 1st floor joists,	2 x 7 "	12	"	420
19 2d floor joists,	2 x 7 "	22	"	486
19 3d "	2 x 6 "	22	"	418
6 posts,	4 x 7 "	20	"	280
2 plates,	4 x 6 "	24	"	98
70 studs,	2 x 4 "	20	"	933
30 gable & extra studs	2 x 4 "	12	"	240
26 rafters,	2 x 7 "	15	"	455
collar girts,	2 x 6 "		"	108

                    Total number of feet, ——— 4002
Enclosing and under floor boards, 4414 feet,
            Top floor boards, 1584 feet.
Shingles, 6¼ M. Clapboards, 2525 feet.

BALLOON FRAMES.

No. 94 building. 22 by 26 feet, posts 12 feet in height.

				Feet.
3 sills,	7 x 7 inches,	22 feet long,	270
6 short sills,	7 x 7 "	13 "	318
30 1st floor joists,	2 x 7 "	13 "	455
20 2nd " "	2 x 7 "	20 "	512
6 posts,	4 x 7 "	12 "	168
plates,	4 x 6 "	52 "	106
104 studs,	2 x 4 "	12 "	832
28 rafters,	2 x 7 "	15 "	490
collar girts,	2 x 6 "		116

Total number of feet, ———— 3267

Enclosing and under floor boards, 3346 feet.
Shingles, 6¾ M. Clapboards, 1665 feet.

No. 95 building, 22 by 26 feet, posts 14 feet in height.
Same as No. 94, posts and studding 14 feet long.
Total number of feet, 3433

Enclosing and under floor boards, 3548 feet.
Shingles, 6¾ M. Clapboards, 1917 feet.

No. 96 building. 22 by 26 feet, posts 20 feet in height.

				Feet.
3 sills,	7 x 7 inches,	22 feet long,	270
6 sills,	7 x 7 "	13 "	318
30 1st floor joists	2 x 7 "	13 "	455
20 2d floor joists	2 x 7 "	22 "	512
21 3d " "	2 x 6 "	22 "	462
6 posts,	4 x 7 "	20 "	280
2 plates,	4 x 6 "	26 "	106
74 studs,	2 x 4 "	20 "	986
30 gable & extra studs,	2 x 4 "	12 "	240
28 rafters,	2 x 7 "	15 "	490
collar girts,	2 x 6 "	about	116

Total number of feet, ————4235

Enclosing and under floor boards, 4700 feet.
Top floor boards, 1716 feet.
Shingles, 6¾ M. Clapboards, 2637 feet.

BALLOON FRAMES.

No. 97 building, 22 by 28 feet, posts 12 feet in height.

					feet.
3 sills,	7 x 7 inches,	22 feet long,	270
6 short sills,	7 x 7	"	14	"	343
30 1st floor joists,	2 x 7	"	14	"	490
22 2nd "	2 x 7	"	22	"	563
6 posts,	4 x 7	"	12	"	168
plates,	4 x 6	"	56	"	112
107 studs,	2 x 4	"	12	"	856
30 rafters,	2 x 7	"	15	"	525
collar girts,	2 x 6				125

Total number of feet, —— 3452

Enclosing and under floor boards, 3562 feet
Top floor boards, 1232 feet.
Shingles, 7¼ M. Clapboards, 1725 feet.

No. 98 building, 22 by 28 feet, posts 14 feet in height.
Same as No. 97, posts and studding 14 feet long.
Total number of feet, 3622
Enclosing and under floor boards, 3762 feet.
Top floor boards, 1232 feet.
Shingles, 7¼ M. Clapboards, 1975 feet.

BALLOON FRAMES.

No. 99 building, 22 by 28 feet, posts 20 feet in height.

			feet.
3 sills,	7 x 7 inches,	22 feet long,	270
6 sills,	7 x 7 "	14 "	343
30 1st floor joists	2 x 7 "	14 "	490
22 2d. "	2 x 7 "	22 "	563
22 3d. "	2 x 6 "	22 "	484
6 posts,	4 x 7 "	20 "	280
2 plates,	4 x 6 "	28 "	112
77 studs,	2 x 4 "	20 "	1026
30 gable & extra studs,	2 x 4 "	12 "	240
30 rafters,	2 x 7 "	15 "	525
collar girts,	2 x 6		125
	Total number of feet,	———	4458

Enclosing and under floor boards, 5000 feet.
Top floor boards 1848 feet.
Shingles, 7¼ M. Clapboards, 2725 feet.

No. 100 building. 22 by 30 feet, posts 12 feet in height

			feet.
3 sills,	7 x 7 inches,	22 feet long,	270
6 short sills	7 x 7 "	15 "	367
30 1st floor joists	2 x 7 "	15 "	525
23 2d. "	2 x 7 "	22 "	589
6 posts,	4 x 7 "	12 "	168
plates,	4 x 6 "	60 "	120
110 studs,	2 x 4 "	12 "	880
32 rafters,	2 x 7 "	15 "	560
collar girts	2 x 6		133
	Total number of feet,	———	3612

Enclosing and under floor boards, 3738 feet.
Top floor boards, 1320 "
Shingles 7⅝ M. Clapboards, 1784 feet.

BALLOON FRAMES.

No. 101 building, 22 by 30 feet, posts 14 feet in height.
 Same as No. 100, with the additional length of posts and studding.
Total number of feet, 3786.
Enclosing and under floor boards, 3946 feet.
Top floor boards, 1320 feet.
Shingles. 7⅝ M. Clapboards, 2045 feet.

No. 102 building, 22 by 30 feet, posts 20 feet in height.

				Feet.
3 sills,	7 x 7 inches,	22 feet long,		270
6 sills,	7 x 7 "	15 "		367
30 first floor joists,	2 x 7 "	15 "		525
23 second floor joists,	2 x 7 "	22 "		589
24 third floor joists,	2 x 6 "	22 "		528
6 posts,	4 x 7 "	20 "		280
2 plates,	4 x 6 "	30 "		120
80 studs,	2 x 4 "	20 "		1066
30 gable and extra studs	2 x 4 "	12 "		240
32 rafters,	2 x 7 "	15 "		560
collar girts,	2 x 6 "			133
		Total number of feet,	——	4678

Enclosing and under floor boards, 5230 feet.
Top floor boards, 1980 feet.
Shingles, 7⅝ M. Clapboards, 2825 feet.

ONE-HALF BALLOON FRAMES.

No. 103 building. 24 by 24 feet, posts 12 feet in height

				feet.
3 sills,	7 x 7 inches,	24 feet long,		294
6 short sills	7 x 7 "	12 "		294
1 tie girt,	7 x 7 "	24 "		98
32 1st floor joists	2 x 7 "	12 "		448
38 2d. "	2 x 7 "	12 "		532
6 posts,	4 x 7 "	12 "		168
plates,	4 x 6 "	48 "		96
105 studs,	2 x 4 "	12 "		840
26 rafters,	2 x 7 "	16½ "		503
13 collar girts	2 x 6			130
		Total number of feet,	——	3403

Enclosing and under floor boards, 3410 feet.
Top floor boards, 1152 "
Shingles 6⅞ M. Clapboards, 1712 feet.

ONE-HALF BALLOON FRAMES.

No. 104 building, 24 by 24 feet, posts 14 feet in height.

Same as No. 103, with the additional length of posts and studding.

Total number of feet, 3560.

Enclosing and under floor boards, 3602 feet.

Top floor boards, 1152 "

Shingles 6⅞ M. Clapboards, 1950 feet.

No. 105 building 24 by 24 feet, posts 20 feet in height.

					Feet.
3 sills,	7 x 7 inches,	24 feet long,			294
6 "	7 x 7 "	12 " "			294
1 tie girt,	7 x 7 "	24 " "			98
32 floor joists,	2 " 7 "	12 " "			448
38 2d " "	2 x 7 "	12 " "			532
19 3d " "	2 x 6 "	24 " "			456
6 posts,	4 " 7 "	20 " "			280
2 plates,	4 x 6 "	24 " "			96
73 studs,	2 " 4 "	20 " "			973
32 gable & extra studs	2 x 4 "	12 " "			256
26 rafters,	2 " 7 "	16½ " "			503
3 collar girts,	2 x 6 "				130

Total number of feet, —— 4360

Enclosing and under floor boards, 4754 feet.

Top floor boards, 1728 "

Shingles, 6⅞ M. Clapboards, 2670 feet.

ONE-HALF BALLOON FRAMES.

No. 106 building, 24 by 26 feet, posts 12 feet in height.

				Feet.
3 sills,	7 x 7 inches,	24 feet long,		294
6 short sills,	7 x 7 "	13 "		318
1 tie girt	7 x 7 "	24 "		98
32 1st floor joists	2 x 7 "	13 "		485
38 2d. "	2 x 7 "	13 "		576
6 posts,	4 x 7 "	12 "		168
plates	4 x 6 "	52 "		104
108 studs,	2 x 4 "	12 "		864
28 rafters,	2 x 7 "	16½ "		539
collar girts,	2 x 6 "			138

Total number of feet, —— 3584

Enclosing and under floor boards, 3620 feet.
Top floor boards, 1248 feet.
Shingles, about 7¾ M. Clapboards, 1770 feet.

No. 107 building, 24 by 26 feet, posts 14 feet in height.

Same as No. 106, with the additional length of posts and studding.

Total number of feet, 3756

Enclosing and under floor boards, 3820 feet.
Top floor boards, 1248 feet.
Shingles, M.7¾ Clapboards, 2020 feet.

ONE-HALE BALLOON FRAMES

No. 108 building, 24 by 26 feet, posts 20 feet in height.

				Feet.
3 sills,	7 x 7 inches,	24 feet long,		294
6 sills,	7 x 7 "	13	"	318
1 tie girt,	7 x 7 "	24	"	98
32 first floor joists,	2 x 7 inches,	13 feet long,		485
38 second "	" 2 x 7 "	13	"	576
21 third "	" 2 x 6 "	24	"	504
6 posts,	4 x 7 "	.20	"	280
2 plates,	4 x 6 "	26	"	106
76 studs,	2 x 4 "	20	"	1013
32 gable and extra studs	2 x 4 "	12	"	256
28 rafters,	2 x 7 "	16½	"	539
14 collar girts,	2 x 6 "			138

Total number of feet, ——— 4607

Enclosing and under floor boards, 5050 feet.
Top floor boards, 1872 feet.
Shingles, 7⅜ M. Clapboards, 2770 feet.

No. 109 building, 24 by 28 feet, posts 12 feet in height.

				Feet.
3 sills,	7 x 7 inches,	24 feet long,		294
6 short sills,	7 x 7 "	14	"	340
1 tie girt,	7 x 7 "	24	"	98
32 first floor joists,	2 x 7 "	14	"	522
38 second floor "	2 x 7 "	14	"	620
6 posts,	4 x 7 "	12	"	168
plates,	4 x 6 "	56	"	112
112 studs,	2 x 4 "	12	"	896
30 rafters,	2 x 7 "	16½	"	577
collar girts,	2 x 6 "			146

Total number of feet, ——— 3773

Enclosing and under floor boards, 3830 feet.
Top floor boards, 1344 feet.
Shingles, 7⅞ M. Clapboards, 1830 feet.

ONE-HALF BALLOON FRAMES.

No 110. building, 24 by 28 feet, posts 14 feet in height.

Same as No. 109 posts and studding 14 feet long.
Total number of feet, 3950
Enclosing and under floor boards, 4038 feet.
Top floor boards, 1344 feet.
Shingles, about 7⅛ M. Clapboards, 2092 feet.

No. 111 building, 24 by 28 feet, posts 20 feet in height.

					Feet.
3 sills,	7 x 7 inches,	24 feet long,			294
6 sills,	7 x 7 "	14 "			340
1 tie girt,	7 x 7 "	24 "			98
32 1st floor joists,	2 x 7 "	14 "			522
38 2d floor joists,	2 x 7 "	14 "			620
22 3d "	2 x 6 "	24 "			528
6 posts,	4 x 7 "	20 "			280
2 plates,	4 x 6 "	28 "			112
80 studs,	2 x 4 "	20 "			1066
32 gable & extra studs	2 x 4 "	12 "			256
30 rafters,	2 x 7 "	16½ "			577
15 collar girts,	2 x 6 "	"			146

Total number of feet, ——— 4839
Enclosing and under floor boards, 5334 feet,
Top floor boards, 2016 feet.
Shingles, 7⅛ M. Clapboards, 2865 feet.

ONE-HALF BALLOON FRAMES.

No. 112 building, 24 by 30 feet, posts 12 feet in height.

				feet.
3 sills,	7 x 7 inches,	24 feet long,		294
6 short sills,	7 x 7 "	15 "		367
1 tie girt,	7 x 7 "	24 "		98
32 1st floor joists	2 x 7 "	15 "		560
38 2d. "	2 x 7 "	15 "		665
6 posts,	4 x 7 "	12 "		168
plates,	4 x 6 "	60 "		120
115 studs,	2 x 4 "	12 "		920
32 rafters,	2 x 7 "	16½ "		616
collar girts,	2 x 6			155

Total number of feet. —— 3963
Enclosing and under floor boards, 4040 feet,
Top floor boards 1440 feet.
Shingles, 8¼ M. Clapboards, 1890 feet.

No. 113 building, 24 by 30 feet, posts 14 feet in height.
Same as No. 112, posts and studding 14 feet long.
Total number of feet, 4144
Enclosing and under floor boards, 4256 feet.
Top floor boards, 1440 feet.
Shingles, 8¼ M. Clapboards, 2160 feet.

ONE-HALF BALLOON FRAMES.

No. 114 building, 24 by 30 feet, posts 20 feet in height.

					Feet.
3 sills,	7 x 7 inches,	24 feet long,	294
6 sills,	7 x 7 "	15 "	367
1 tie girt,	7 x 7 "	24 "	98
32 1st. floor joists,	2 x 7 "	15 "	560
38 2d. "	2 x 7 "	15 "	665
24 3d. floor joists,	2 x 6 "	24 "	576
6 posts,	4 x 7 "	20 "	280
2 plates,	4 x 6 "	30 "	120
82 studs,	2 x 4 "	20 "	1083
32 gable studs,	2 x 4 "	12 "	256
32 rafters,	2 x 7 "	16½ "	616
collar girts,	2 x 6			155

Total number of feet, —— 5070

Enclosing and under floor boards, 5624 feet.
Top floor boards, 2160 "
Shingles, 8½ M. Clapboards, 2970 feet.

No. 115 building, 26 by 26 feet, posts 12 feet in height.

				feet.
3 sills,	7 x 7 inches,	26 feet long,	318
6 short sills,	7 x 7 "	13 "	318
1 tie girt,	7 x 7 "	26 "	106
36 1st floor joists,	2 x 7 "	13 "	546
40 2nd "	2 x 7 "	13 "	606
6 posts,	4 x 7 "	12 "	168
plates,	4 x 6 "	52 "	106
115 studs,	2 x 4 "	12 "	920
28 rafters,	2 x 7 "	17¼ "	573
collar girts,	2 x 7			196

Total number of feet, —— 3857

Enclosing and under floor boards, 3866 feet
Top floor boards, 1352 feet.
Shingles, 7¾ M. Clapboards, 1875 feet.

ONE-HALF BALLOON FRAMES.

No. 116 building, 26 by 26 feet. posts 14 feet in height.

Same as No. 115, with the additional length of posts and studding.

Total number of feet, 4038

Enclosing and under floor boards, 4074 feet.
Top floor boards, 1352 feet.
Shingles, 7⅞ M. Claphoards, 2135 feet.

No. 117 building, 26 by 26 feet, posts 20 feet in height.

					feet.
3 sills,	7 x 7 inches,	26 feet long,	318
6 short sills,	7 x 7	"	13	"	318
1 tie girt,	7 x 7	"	26	"	106
36 1st floor joists,	2 x 7	"	13	"	546
40 2d floor joists,	2 x 7	"	13	"	606
20 3d "	2 x 7	"	26	"	606
6 posts,	4 x 7	"	20	"	280
2 plates,	4 x 6	"	26	"	106
80 studs,	2 x 4	"	20	"	1066
34 gable & extra studs	2 x 4	"	12	"	272
28 rafters,	2 x 7	"	17½	"	573
collar girts,	2 x 7	"		about	196

Total number of feet, —— 4993

Enclosing and under floor boards, 5374 feet.
Top floor boards, 2028 feet.
Shingles, 7⅞ M. Clapboards, 2915 feet.

ONE-HALF BALLOON FRAMES.

No. 118 building, 26 by 28 feet, posts 12 feet in height.

				Feet.
3 sills	7 x 7 inches,	26 feet long,		318
6 short sills,	7 x 7 "	14 "		343
1 tie girt,	7 x 7 "	26 "		106
36 first floor joists,	2 x 7 "	14 "		588
40 second floor joists,	2 x 7 "	14 "		653
6 posts,	4 x 7 "	12 "		168
plates,	4 x 6 "	56 "		112
118 studs,	2 x 4 "	12 "		944
30 rafters,	2 x 7 "	17½ "		615
collar girts,	2 x 7		about	210

Total number of feet, —— 4057

Enclosing and under floor boards, 4088 feet.
Top floor boards, 1456 feet.
Shingles, M. 8¼ Clapboards, 1935 feet.

No. 119 building, 26 by 28 feet, posts 14 feet in height.

Same as No. 118, with the additional length of posts and studding.

Total number of feet, 4242

Enclosing and under floor boards, 4304 feet.
Top floor boards, 1456 feet.
Shingles, 8¼ M. Clapboards, 2200 feet.

ONE-HALF BALLOON FRAMES.

No. 120 building, 26 by 28 feet, posts 20 feet in height.

					Feet.
3 sills,	7 x 7 inches,	26 feet long,	318
6 sills,	7 x 7 "	14 "	343
1 tie girt	7 x 7 "	26 "	106
36 1st floor joists	2 x 7 "	14 "	588
40 2d floor joists,	2 x 7 "	14 "	653
22 3d floor joists,	2 x 7 "	26 "	667
6 posts,	4 x 7 "	20 "	280
2 plates,	4 x 6 "	28 "	112
83 studs,	2 x 4 "	20 "	1079
34 gable and extra studs	2 x 4 "	12 "	272
30 rafters,	2 x 7 "	17½ "	615
collar girts,	2 x 7 "		210
	Total number of feet,		——— 5243

Enclosing and under floor boards, 5678 feet.
Top floor boards, 2184 feet.
Shingles, 8¼ M. Clapboards, 3015 feet.

No. 121 building, 26 by 30 feet, posts 12 feet in height.

				Feet.
3 sills,	7 x 7 inches,	26 feet long,	318
6 short sills,	7 x 7 "	15 "	367
1 tie girt	7 x 7 "	26 "	106
36 1st floor joists,	2 x 7 "	15 "	630
40 2d " "	2 x 7 "	15 "	700
6 posts,	4 x 7 "	12 "	168
plates,	4 x 6 "	60 "	120
121 studs,	2 x 4 "	12 "	968
32 rafters,	2 x 7 "	17½ "	656
16 collar girts,	2 x 7 "		238
	Total number of feet,		——— 4271

Enclosing and under floor boards, 4311 feet.
Top floor boards 1560 feet.
Shingles, 8¼ M. Clapboards, 1997 feet.

ONE-HALF BALLOON FRAMES.

No. 122. building, 26 by 30 feet, posts 14 feet in height.

Same as No. 121 posts and studding 2 feet longer
       Total number of feet, 4460
Enclosing and under floor boards, 4535 feet.
    Top floor boards, 1560 feet.
Shingles, about 8⅞ M. Clapboards, 2276 feet.

No. 123 building, 26 by 30 feet, posts 20 feet in height.

				Feet.
3 sills,	7 x 7 inches,	26 feet long,	318
6 sills,	7 x 7 "	15 "	367
1 tie girt.	7 x 7 "	26 "	106
36 1st floor joists,	2 x 7 "	15 "	630
40 2d floor joists,	2 x 7 "	15 "	700
24 3d " "	2 x 7 "	26 "	728
6 posts,	4 x 7 "	20 "	280
2 plates,	4 x 6 "	30 "	120
86 studs,	2 x 4 "	20 "	1146
34 gable & extra studs	2 x 4 "	12 "	272
32 rafters,	2 x 7 "	17½ "	656
16 collar girts,	2 x 7 "	"	238

      Total number of feet, —— 5561
Enclosing and under floor boards, 5989 feet.
   Top floor boards, 2340 feet.
Shingles, 8⅞ M. Clapboards, 3116 feet.

ONE-HALF BALLOON FRAMES.

No. 124 building, 26 by 32 feet, posts 12 feet in height.

				feet.
4 sills,	7 x 7 inches,	26 feet long,	424
6 sills,	7 x 7 "	13 "	318
3 short sills,	7 x 7 "	6 "	73
2 tie girts,	7 x 7 "	26 "	212
36 1st floor joists,	2 x 7 "	13 "	546
18 1st floor joists,	2 x 6 "	6 "	126
40 2d floor joists,	2 x 7 "	13 "	606
20 2d floor joists,	2 x 7 "	6 "	140
8 posts,	4 x 7 "	12 "	224
2 plates,	4 x 6 "	64 "	128
125 studs,	2 x 4 "	12 "	1000
34 rafters,	2 x 7 "	17½ "	697
17 collar girts,	2 x 7 "	· about	250
		Total number of feet,	—— 4744

Enclosing and under floor boards, 4530 feet.
        Top floor boards, 1664 feet.
Shingles, 9½ M   Clapboards, 2055 feet.

———

No. 125 building, 26 by 32 feet, posts 14 feet in height.
  Same as No. 124, posts and studding 2 feet longer.
               Total number of feet,   4947
Enclosing and under floor boards, 4762 feet.
        Top floor boards, 1664 feet.
Shingles, 9½ M.   Clapboards, 2345 feet.

ONE-HALF BALLOON FRAMES

No. 126 building, 26 by 32 feet, posts 20 feet in height.

					Feet.	
4 sills,	7 x 7 inches,	26 feet long,	424	
6 sills,	7 x 7 "	13 "	318
3 short sills,	7 x 7 "	6 "	73
2 tie girts,	7 x 7 "	26 "	212
36 first floor joists,	2 x 7 "	13 "	546
18 first floor joists	2 x 7 "	6 "	126
40 second " "	2 x 7 "	13 "	606
20 second floor joists	2 x 7 "	6 "	140
25 third " "	2 x 7 "	26 "	758
8 posts,	4 x 7 "	20 "	340
2 plates,	4 x 6 "	32 "	128
90 studs,	2 x 4 "	20 "	1200
34 gable and extra studs	2 x 4 "	12 "	272
34 rafters,	2 x 7 "	17½ "	697
17 collar girts,	2 x 7 "		250

Total number of feet, ——— 6090

Enclosing and under floor boards, 6300 feet.
Top floor boards, 2496 feet.
Shingles, 9½ M. Clapboards, 3215 feet.

No. 127 building, 26 by 34 feet, posts 12 feet in height.

					Feet.
4 sills,	7 x 7 inches,	26 feet long,	424
6 sills,	7 x 7 "	13 "	318
3 short sills	7 x 7 "	8 "	98
2 tie girts,	7 x 7 "	26 "	212
36 first floor joists,	2 x 7 "	13 "	546
18 first floor joists	2 x 7 "	8 "	168
40 second floor "	2 x 7 "	13 "	606
20 second floor joists	2 x 7 "	8 "	186
8 posts,	4 x 7 "	12 "	224
plates,	4 x 6 "	68 "	136
128 studs,	2 x 4 "	12 "	1024
36 rafters,	2 x 7 "	17½ "	738
18 collar girts,			250

Total number of feet, ——— 4930

Enclosing and under floor boards, 4754 feet.
Top floor boards. 1768 feet.
Shingles, 10 M. Clapboards, 2115 feet.
No. 127 in full frame girts and plates 7 x 7 inches 5550 feet.

ONE HALF BALLOON FRAMES.

No. 128 building, 26 by 34 feet, posts 14 feet in height. Same as No. 127, posts and studding 14 feet long.

Total number of feet, . 5128

Enclosing and under floor boards, 4994 feet.

Top floor boards, 1768 feet.

Shingles, 10 M. Clapboards, 2415 feet.

No. 129 building. 26 by 34 feet, posts 20 feet in height.

					Feet.
4 sills,	7 x 7 inches,	26 feet long,			424
6 sills,	7 x 7	"	13	"	318
3 short sills,	7 x 7	"	8	"	98
2 tie girts	7 x 7	"	26	"	212
36 1st floor joists	2 x 7	"	13	"	546
18 1st floor joists	2 x 7	"	8	"	168
40 2d floor joists	2 x 7	"	13	"	606
20 2d floor joists	2 x 7	"	8	"	186
27 3d " "	2 x 7	"	26	"	819
8 posts,	4 x 7	"	20	"	340
2 plates,	4 x 6	"	34	"	136
93 studs,	2 x 4	"	20	"	1140
34 gable & extra studs,	2 x 4	"	12	"	272
36 rafters,	2 x 7	"	17½	"	738
18 collar girts,	2 x 7		about		250

Total number of feet, ———6253

Enclosing and under floor boards, 6600 feet.

Top floor boards, 2652 feet.

Shingles, 10 M. Clapboards, 3315 feet.

Full frame No. 129, girts and plates 7 x 7 about 7500.

ONE-HALF BALLOON FRAMES.

No. 130 building, 26 by 36 feet, posts 12 feet in height.

						feet.
4 sills,	7 x 7	inches,	26	feet long,		424
6 "	7 x 7	"	14	"		343
3 short sills,	7 x 7	"	8	"		98
2 tie girts,	7 x 7	"	26	"		212
36 1st floor joists	2 x 7	"	14	"		588
18 1st floor joists,	2 x 7	"	8	"		168
20 2d, "	2 x 7	"	8	"		186
40 2d " "	2 x 7	"	14	"		652
8 posts,	4 x 7	"	12	"		224
plates,	4 x 6	"	72	"		144
130 studs,	2 x 4	"	12	"		1040
38 rafters,	2 x 7	"	17½	"		779
collar girts,	2 x 7					265
		Total number of feet.			——	5123

Enclosing and under floor boards, 4976 feet,
　　　Top floor boards 1872 feet.
Shingles, 10½ M. Clapboards, 2175 feet.
Full frame No. 130 girts and plates 7 x 7 in. about 5800

No. 131 building, 26 by 36 feet, posts 14 feet in height.
　Same as No. 130, posts and studding 14 feet long.
　　　　　　　Total number of feet, 5333
Enclosing and under floor boards, 5224 feet.
　　　Top floor boards, 1872 feet.
Shingles, 10½ M. Clapboards, 2485 feet.

ONE-HALF BALLOON FRAMES.

No. 132 building. 26 by 36 feet, posts 20 feet in height

				feet.
4 sills,	7 x 7 inches,	26 feet long,		424
6 sills	7 x 7	"	14 "	343
3 short sills,	7 x 7	"	8 "	98
2 tie girts,	7 x 7	"	26 "	212
36 1st floor joists	2 x 7	"	14 "	588
18 first floor joists,	2 x 7	"	8 "	168
40 2d. "	2 x 7	"	14 "	652
20 second floor joists,	2 x 7	"	8 "	186
28 third floor joists,	2 x 7	"	26 "	849
8 posts,	4 x 7	"	20 "	340
2 plates,	4 x 6	"	36 "	144
96 studs,	2 x 4	"	20 "	1280
34 gable & extra studs	2 x 4	"	12 "	272
38 rafters,	2 x 7	"	17½ "	779
19 collar girts,	2 x 7			265

Total number of feet, —— 6600

Enclosing and under floor boards, 6900 feet.
Top floor boards, 2808 "
Shingles 10½ M. Clapboards. 3415 feet.
Full frame No. 132 girts and plates 7 x 7 in. about 7750 feet.

No. 133 building 26 by 38 feet, posts 12 feet in height.

				Feet.
4 sills,	7 x 7	inches,	26 feet long,	424
6 "	7 x 7	"	14 "	343
3 short sills,	7 x 7	"	10 "	123
2 tie girts,	7 x 7	"	26 "	212
36 1st floor joists,	2 x 7	"	14 "	588
18 1st floor joists,	2 x 7	"	10 "	210
20 2d. "	2 x 7	"	10 "	232
40 2d. floor joists,	2 x 7	"	14 "	653
8 posts,	4 x 7	"	12 "	224
plates,	4 x 6	"	76 "	152
134 studs,	2 x 4	"	12 "	1072
40 rafters,	2 x 7	"	17½ "	820
20 collar girts,	2 x 7	"		280

Total number of feet, —— 5333

Enclosing and under floor boards, 5198 feet.
Top floor boards. 1976 "
Shingles, 11 M. Clapboards, 2235 feet.
Full frame No. 133, all girts and plates 7 x 7 in. 5900 feet.

ONE-HALF BALLOON FRAMES.

No. 134 building, 26 by 38 feet, posts 14 feet in height.

Same as No. 133, posts and studding 14 feet long.
Total number of feet,   5548
Enclosing and under floor boards, 5454 feet.
Top floor boards, 1926 feet.
Shingles, 11 M. Clapboards, 2555 feet.

No. 135 building, 26 by 38 feet, posts 20 feet in height.

					Feet.
4 sills,	7 x 7 inches,	26 feet long,	424
6 "	7 x 7 "	14 "	343
3 short sills,	7 x 7 "	10 "	123
2 tie girts	7 x 7 "	26 "	212
36 1st floor joists	2 x 7 "	14 "	588
18 1st floor joists	2 x 7 "	10 "	210
40 2d. "	2 x 7 "	14 "	653
20 2 d. "	2 x 7 "	10 "	232
30 3d "	2 x 7 "	26 "	910
8 posts,	4 x 7 "	20 "	340
2 plates	4 x 6 "	38 "	152
100 studs,	2 x 4 "	12 "	1333
34 gable & extra studs	2 x 4 "	12 "	272
40 rafters,	2 x 7 "	17½ "	820
20 collar girts,	2 x 7 "		280

Total number of feet,   ——— 6892
Enclosing and under floor boards, 7200 feet.
Top floor boards, 2964 feet.
Shingles, 11 M. Clapboards, 3515 feet.
Full frame No. 135 full girts and plates about 7950 feet

ONE HALF BALLOON FRAMES

No. 136 building. 26 by 40 feet, posts 12 feet in height.

				Feet.
4 sills,	7 x 7 inches,	26 feet long,	424
6 sills,	7 x 7 "	15 "	367
3 "	7 x 7 "	10 "	123
2 tie girts,	7 x 7 "	26 "	212
36 1st. floor joists,	2 x 7 "	15 "	630
18 1st. " "	2 x 7 "	10 "	210
20 2d. "	2 x 7 "	10 "	232
40 2d. "	2 x 7 "	15 "	700
8 posts,	4 x 7 "	12 "	224
plates,	4 x 6 "	80 "	160
138 studs,	2 x 4 "	12 "	1104
42 rafters,	2 x 7 "	17½ "	861
21 collar girts,	2 x 7	12	294
	Total number of feet,	——	5541

Enclosing and under floor boards, 5420 feet.
Top floor boards, 2080 "
Shingles, 11⅝ M. Clapboards, 2295 feet.
Full frames No. 136 girts and plates 7 x 7 about 6200 feet.

No. 137 building, 26 by 40 feet, posts 14 feet in height.

Same as No. 136, posts and studding 14 feet long.
Total number of feet, 5762.
Enclosing and under floor boards, 5684 feet.
Shingles 11⅝ M. Clapboards, 2625 feet.

ONE-HALF BALLOON FRAMES

No. 138 building, 26 by 40 feet, posts 20 feet in height.

				Feet.
4 sills,	7 x 7 inches.	26 feet long,	424
6 sills,	7 x 7 "	15 "	367
3 short sills,	7 x 7 "	10 "	123
2 tie girts,	7 x 7 "	26 "	212
36 first floor joists,	2 x 7 "	15 "	630
18 first floor joists	2 x 7 "	10 "	210
40 second " "	2 x 7 "	15 "	700
20 second floor joists	2 x 7 "	10 "	232
31 third " "	2 x 7 "	26 "	940
8 posts,	4 x 7 "	20 "	340
2 plates,	4 x 6 "	40 "	160
104 studs,	2 x 4 "	20 "	1386
34 gable and extra studs	2 x 4 "	12 "	272
42 rafters,	2 x 7 "	17½ "	861
21 collar girts,	2 x 7 "	12 "	294
	Total number of feet,		7151

Enclosing and under floor boards, 7472 feet.
Top floor boards, 3120 feet.
Shingles, 11⅜ M. Clapboards, 3615 feet.
Full frame No. 130, girts and plates 7 x 7 in. about 8300 feet.

ONE-HALF BALLOON FRAMES.

No. 139 building, 28 by 28 feet, posts 12 feet in height.

					feet.
3 sills,	7 x 7 inches,	28 feet long,	343
6 short sills,	7 x 7	"	14	"	343
1 tie girt,	7 x 7	"	28	"	114
38 1st floor joists,	2 x 7	"	14	"	620
44 2d floor joists,	2 x 7	"	14	"	718
8 posts,	4 x 7	"	12	"	224
plates,	4 x 6	"	56	"	112
125 studs,	2 x 4	"	12	"	1000
30 rafters,	2 x 7	"	19	"	665
collar girts,	2 x 7	"			245

Total number of feet, —— 4384

Enclosing and under floor boards, 4390 feet.
Top floor boards, 1568 feet.
Shingles, 9 M. Clapboards, 2055 feet.
Full frame No. 139 girt and plates, 7 x 7 about 4900 feet.

No. 140 building, 28 by 28 feet, posts 14 feet in height.
Same as No. 139, posts and studding 14 feet long.
Total number of feet, 4587
Enclosing and under floor boards, 4640 feet.
Top floor boards, 1568 feet.
Shingles, 9 M. Clapboards, 2335 feet.

ONE-HALF BALLOON FRAMES.

No. 141 building, 28 by 28 feet, posts 20 feet in height.

				Feet.
3 sills	7 x 7 inches,	28 feet long,	343
6 sills,	7 x 7 "	14 "	343
1 tie girt,	7 x 7 "	28 "	114
38 first floor joists,	2 x 7 "	14 "	620
44 second floor joists,	2 x 7 "	14 "	718
22 third floor joists	2 x 7 "	28 "	718
8 posts,	4 x 7 "	20 "	340
2 plates,	4 x 6 "	28 "	112
86 studs,	2 x 4 "	20 "	1147
36 gable & extra studs	2 x 4 "	12 "	288
30 rafters,	2 x 7 "	19 "	665
15 collar girts,	2 x 7 "	14 "	245

Total number of feet, ——— 5653

Enclosing and under floor boards, 6070 feet.
Top floor boards, 2352 feet.
Shingles, 9 M. Clapboards, 3172 feet.
Full frames No. 141 girts and plates 7 x 7 about 6600

No. 142 building, 28 by 30 feet, posts 12 feet in height.

				Feet.
3 sills,	7 x 7 inches,	28 feet long,	343
6 sills,	7 x 7 "	15 "	367
1 tie girts,	7 x 7 "	28 "	114
38 first floor joists,	2 x 7 "	15 "	665
44 second floor "	2 x 7 "	15 "	770
8 posts,	4 x 7 "	12 "	224
plates,	4 x 6 "	60 "	120
130 studs,	2 x 4 "	12 "	1040
32 rafters,	2 x 7 "	19 "	709
collar girts	2 x 7	14	261

Total number of feet, ——— 4613

Enclosing and under floor boards, 4626 feet.
Top floor boards, 1680 feet.
Shingles, 9¾ M. Clapboards, 2115 feet.
Full frame No. 142, girts and plates 7 x 7 inches 5200 feet.

ONE-HALF BALLOON FRAMES.

No. 143 building, 28 by 30 feet, posts 14 feet in height. Same as No. 142, posts and studding 14 feet long.
 Total number of feet, 4821
Enclosing and under floor boards, 4858 feet.
 Top floor boards, 1680 feet.
Shingles, 9¾ M. Clapboards, 2500 feet.

No. 144 building, 28 by 30 feet, posts 20 feet in height.

					feet.
3 sills,	7 x 7 inches,	28 feet long,			343
6 sills	7 x 7	"	15	"	367
1 tie girt,	7 x 7	"	28	"	114
38 1st floor joists	2 x 7	"	15	"	665
44 2d. "	2 x 7	"	15	"	770
23 3d. floor joists,	2 x 7	"	28	"	741
8 posts,	4 x 7	"	20	"	340
2 plates,	4 x 6	"	30	"	120
90 studs,	2 x 4	"	20	"	1200
36 gable & extra studs	2 x 4	"	12	"	288
32 rafters,	2 x 7	"	19	"	709
16 collar girts,	2 x 7	"	14	"	261

 Total number of feet, —— 5918
Enclosing and under floor boards, 6398 feet.
 Top floor boards, 2520 "
Shingles 9¾ M. Clapboards, 3280 feet.
Full frame No. 144 girts and plates 7 x 7 in. about 7000 feet.

ONE-HALF BALLOON FRAMES.

No. 145 building 28 by 32 feet, posts 12 feet in height.

				Feet.
4 sills,	7 x 7 inches,	28 feet long,	457
6 "	7 x 7 "	12 "	354
3 short sills,	7 x 7 "	8 "	98
2 tie girts,	7 x 7 "	28 "	228
38 1st floor joists,	2 x 7 "	12 "	532
19 1st floor joists,	2 x 7 "	8 "	177
22 2d. "	2 x 7 "	8 "	205
44 2d. floor joists,	2 x 7 "	12 "	616
8 posts,	4 x 7 "	12 "	224
plates,	4 x 6 "	64 "	128
134 studs,	2 x 4 "	12 "	1072
34 rafters,	2 x 7 "	19 "	753
16 collar girts,	2 x 7 "		277

Total number of feet, ——— 5121

Enclosing and under floor boards, 4852 feet.
Top floor boards, 1792 "
Shingles, 10¼ M. Clapboards, 2175 feet.
Full frame No. 145, all girts and plates 7 x 7 in. 5600 feet.

No. 146 building, 28 by 32 feet, posts 14 feet in height.
Same as No. 145, posts and studding 14 feet long.

Total number of feet, 5336

Enclosing and under floor boards, 5092 feet.
Top floor boards, 1792 feet.
Shingles, 10¼ M. Clapboards, 2475 feet.

ONE HALF BALLOON FRAMES.

No. 147 building, 28 by 32 feet, posts 20 feet in height.

				Feet.
4 sills,	7 x 7 inches,	28 feet long,	457
6 sills,	7 x 7 "	12 "	354
3 short sills,	7 x 7 "	8 "	98
2 tie girts	7 x 7 "	28 "	228
38 1st floor joists	2 x 7 "	12 "	532
19 1st floor joists	2 x 7 "	8 "	177
44 2d floor joists	2 x 7 "	12 "	616
22 2d floor joists	2 x 7 "	8 "	205
25 3d " "	2 x 7 "	28 "	816
8 posts,	4 x 7 "	20 "	340
2 plates,	4 x 6 "	32 "	128
97 studs.	2 x 4 "	20 "	1293
30 gable studs,	2 x 4 "	12 "	240
34 rafters,	2 x 7 "	19 "	753
16 collar girts,	2 x 7		277
	Total number of feet,		———6514

Enclosing and under floor boards, 6708 feet.
Top floor boards, 2688 feet.
Shingles. 10¼ M. Clapboards, 3375 feet.
Full frame No. 147, girts and plates 7 x 7 about 7600 feet.

ONE-HALF BALLOON FRAMES.

No. 148 building, 28 by 34 feet, posts 16 feet in height.

				feet.
4 sills,	7 x 7 inches,	28 feet long,		457
6 "	7 x 7 "	13 "		318
3 short sills,	7 x 7 "	8 "		98
2 tie girts,	7 x 7 "	28 "		228
38 1st floor joists	2 x 7 "	13 "		576
19 1st floor joists,	2 x 7 "	8 "		177
22 2d, "	2 x 7 "	8 "		205
44 2d " " .	2 x 7 "	13 "		667
26 3d " .'	2 x 7 "	28 "		848
8 posts,	4 x 7 "	16 "		266
plates,	6 x 6 "	68 "		204
135 studs,	2 x 4 "	16 "		1438
36 rafters,	2 x 7 "	19 "		798
18 collar girts,	2 x 7 "	14 "		294

Total number of feet. —— 6574

Enclosing and under floor boards, 5594 feet,
Top floor boards 2856 feet.
Shingles, 10⅞ M. Clapboards, 2835 feet.
Full frame No. 148 girts and plates, about 7650

No. 149 building, 28 by 34 feet, posts 20 feet in height.

Same as No. 148 with the additional length of posts and studding.

Total number of feet, 7000
Enclosing and under floor boards, 7092 feet.
Top floor boards, 2856 feet.
Shingles, 10⅞ M. Clapboards, 3475 feet.
Full frame No. 149 about 8000

ONE-HALF BALLOON FRAMES.

No. 150 building, 28 by 34 feet, posts 12 feet in height.

				Feet.
4 sills,	7 x 7 inches,	28 feet long,		457
6 sills,	7 x 7 "	13 "		318
3 short sills,	7 x 7 "	8 "		98
2 tie girts,	7 x 7 "	28 "		228
38 1st floor joists,	2 x 7 "	13 "		578
19 1st floor joists,	2 x 7 "	8 "		177
22 2d floor joists,	2 x 7 "	8 "		205
44 2d floor joists,	2 x 7 "	13 "		667
8 posts,	4 x 7 "	12 "		224
2 plates,	6 x 6 "	34 "		204
135 studs,	2 x 4 "	12 "		1080
36 rafters,	2 x 7 "	19 "		798
18 collar girts,	2 x 7 "	14 "		294

Total number of feet, —— 5328

Enclosing and under floor boards, 5098 feet,
Top floor boards, 1904 feet.
Shingles, 10⅞ M. Clapboards, 2235 feet.

FULL FRAMES.

No. 151 building, 30 by 30 feet, posts 18 feet in height.

					Feet.
4 sills,	7 x 7 inches,	30	feet long,	490
6 sills,	7 x 7 "	12	"	294
3 short sills,	7 x 7 "	6	"	73
2 tie girts,	7 x 7 "	30	"	245
4 end girts,	7 x 7 "	15	"	245
4 side girts,	6 x 7 "	12	"	168
2 side girts,	6 x 7 "	6	"	42
4 beams	7 x 7 "	30	"	490
4 plates,	7 x 7 "	12	"	196
2 plates	7 x 7 "	6	"	49
10 posts,	4 x 7 "	18	"	420
40 1st floor joists,	2 x 7 "	12	"	560
20 1st floor joists	2 x 7 "	6	"	140
46 2d floor joists,	2 x 7 "	12	"	644
23 2d floor joists	2 x 7 "	6	"	161
46 3d floor joists,	2 x 7 "	12	"	644
23 3d floor joists	2 x 7 "	6	"	161
90 1st floor studs	2 x 4 "	9½	"	570
85 2d floor studs	2 x 4 "	8½	"	481
45 gable & extra studs,	2 x 4 "	12	"	360
32 rafters,	2 x 7 "	20¼	"	768
16 collar girts,	2 x 7 "	14	"	261

Total number of feet, ——— 7462

Enclosing and under floor boards, 6552 feet.
Top floor boards, 2700 feet.
Shingles. 10½ M. Clapboards, 3125 feet.

FULL FRAMES.

No. 152 building, 30 by 30 feet, posts 22 feet in height.

Same as No. 151, with the additional length of posts and studding, (use long plates.)
Total number of feet, 7793
Enclosing and under floor boards, 7032 feet.
Top floor boards, 2700 feet.
Shingles, 10½ M. Clapboards, 3725 feet.

No. 153 building, 30 by 30 feet, posts 12 feet in height.

Same as No. 151   Number of Feet. 7462
                                                      Feet.
Less 4 beams   7 x 7 inches 30 feet long   490
 " 46 floor joists 2 x 7   "   12   "   644
 " 23 floor joists 2 x 7   "   6   "   161
 " Length of studding   331
 " Length of posts   140
                                                ——— 1724

Total number of feet, 5738

Enclosing and under floor boards, 4932 feet.
Top floor boards, 1800 feet.
Shingles, 10½ M. Clapboards, 2225 feet.

FULL FRAMES.

No. 154 building, 30 by 32 feet, posts 18 feet in height.

				Feet.
4 sills	7 x 7 inches,	30 feet long,		490
6 sills,	7 x 7 "	13 "		318
3 short sills,	7 x 7 "	6 "		73
2 tie girts,	7 x 7 "	30 "		245
4 end girts	7 x 7 "	15 "		245
4 side girts	6 x 7 "	13 "		182
2 side girts	6 x 7 "	6 "		42
4 beams	7 x 7 "	30 "		490
4 plates,	7 x 7 "	13 "		212
2 plates,	7 x 7 "	6 "		49
10 posts,	4 x 7 "	18 "		420
40 first floor joists,	2 x 7 "	13 "		606
20 1st floor joists,	2 x 7 "	6 "		140
46 second floor joists,	2 x 7 "	13 "		697
23 2d " "	2 x 7 "	6 "		161
46 third floor joists	2 x 7 "	13 "		697
23 3d floor joists	2 x 7 "	6 "		161
94 1st floor studs,	2 x 4 "	$9\tfrac{1}{2}$ "		594
89 2d " studs,	2 x 4 "	$8\tfrac{1}{2}$ "		503
45 gable & extra studs	2 x 4 "	12 "		360
34 rafters,	2 x 7 "	$20\tfrac{1}{2}$ "		816
17 collar girts,	2 x 7 "	14 "		278

Total number of feet, ——7779

Enclosing and under floor boards, 6888 feet.
Top floor boards, 2880 feet.
Shingles, 11 M. Clapboards, 3215 feet.

FULL FRAMES

No. 155 building, 30 by 32 feet, posts 22 feet in height.

Same as No. 154, with the adcitioual length of psts and studding,—use long plates.
Total number of feet, 8122
Enclosing and under floor boards, 7382 feet.
Top floor boards, 2880 feet.
Shingles, 11 M.   Clapboards, 3832 feet.

No. 156 building, 30 by 32 feet, posts 12 feet in height

Same as No. 154,
Less 4 beams and upper floor joists.
" Length of posts and studding.
Use long plates,   Total   5946
Enclosing and under floor boards, 5184 feet.
Top floor boards, 1920 feet.
Shingles, 11 M.   Clapboards, 2285 feet.

FULL FRAMES.

No. 157 building, 30 by 34 feet, posts 18 feet in height.

				Feet.
4 sills,	7 x 7 inches,	30	feet long,	490
6 sills,	7 x 7 "	13	"	318
3 short sills,	7 x 7 "	8	"	98
2 tie girts,	7 x 7 "	30	"	245
4 end girts,	7 x 7 "	15	"	245
4 side girts	6 x 7 "	13	"	182
2 side girts,	6 x 7 "	8	"	56
4 beams,	7 x 7 "	30	"	490
4 plates,	7 x 7 "	13	"	212
2 plates,	7 x 7 "	8	"	65
10 posts,	4 x 7 "	18	"	420
40 1st floor joists	2 x 7 "	13	"	606
20 1st floor joists	2 x 7 "	8	"	186
46 2d floor joists	2 x 7 "	13	"	697
23 2d floor joists	2 x 7 "	8	"	214
46 3d " "	2 x 7 "	13	"	697
23 3d " "	2 x 7 "	8	"	214
98 1st floor studs,	2 x 4 "	9½	"	621
93 2d " studs,	2 x 4 "	8½	"	527
45 gable & extra studs,	2 x 4 "	12	"	360
36 rafters,	2 x 7 "	20½	"	864
18 collar girts,	2 x 7	14	"	294
	Total number of feet,			———8101

Enclosing and under floor boards, 7224 feet.
Top floor boards, 3060 feet.
Shingles, 11¾ M. Clapboards, 3305 feet.

FULL FRAMES

No. 158 Building, 30 by 34 feet, posts 22 feet in height.

Same as No. 157 with the additional length of posts and studding, use long plates,

Total 8456

Enclosing and under floor boards 7732

Top floor boards 3060

Shingles 11¾ M. Clapboards 3945 feet.

No. 159 Building 30 by 34 feet, posts 12 feet in height.

Same as No. 157.

Less 4 Beams and upper floor joists.
Length of posts and studding.
Use long plates. Total number of feet, 6189

Enclosing and under floor boards 5432 feet
Top floor boards 2040 "
Shingles 11¾ M. Clapboards 2345 feet.

FULL FRAMES

No. 160 building, 30 by 36 feet, posts 18 feet in height.

				Feet.
4 sills,	7 x 7 inches,	30 feet long,	490
6 sills,	7 x 7 "	14 "	343
3 short sills,	7 x 7 "	8 "	98
2 tie girts,	7 x 7 "	30 "	245
4 end girts	7 x 7 "	15 "	245
4 side girts	6 x 7 "	14 "	196
2 side girts	6 x 7 "	8 "	56
4 beams	7 x 7 "	30 "	490
4 plates,	7 x 7 "	14 "	228
2 plates	7 x 7 "	8 "	65
10 posts,	4 x 7 "	18 "	420
40 first floor joists,	2 x 7 "	14 "	653
20 first floor joists	2 x 7 "	8 "	186
46 second " "	2 x 7 "	14 "	751
23 second floor joists	2 x 7 "	8 "	214
46 third " "	2 x 7 "	14 "	751
23 3d. "	2 x 7 "	8 "	214
100 1st floor studs,	2 x 4 "	9½ "	633
95 2d floor studs	2 x 4 "	8½ "	539
45 gable and extra studs	2 x 4 "	12 "	360
38 rafters,	2 x 7 "	20½ "	912
19 collar girts,	2 x 7 "	14 "	310

Total number of feet, —— 8399

Enclosing and under floor boards, 7554 feet.
Top floor boards, 3240 feet.
Shingles, 12¼ M. Clapboards, 3395 feet.

No. 161 building, 30 by 36 feet, posts 22 feet in height.

Same as No. 160, with the additional length of posts and studding.

Use long plates. Total number of feet 8753
Enclosing and under floor boards, 8072 feet.
Top " 3240 feet.
Shingles 12¼ M. Clapboards, 4042 feet.

FULL FRAMES

No. 162 building, 30 by 36 feet, posts 12 feet in height.

    Same as No. 160.
Less 4 beams and upper floor joists.
" 6 feet in length of posts and studding,— use long plates
                Total number of feet,   6535
        Enclosing and under floor boards, 5682
                Top floor boards, 2160
Shingles, 12¼ M.   Clapboards, 2400 feet.

No. 163 building 32 by 32 feet, posts 18 feet in height.

					Feet.
4 sills,	7 x 7	inches,	32	feet long,	522
6 "	7 x 7	"	13	"	318
3 short sills,	7 x 7	"	6	"	73
2 tie girts,	7 x 7	"	32	"	261
4 end girts,	7 x 7	"	16	"	261
4 side girts,	6 x 7	"	13	"	182
2 side girts,	6 x 7	"	6	"	42
4 beams,	7 x 7	"	32	"	522
4 plates,	7 x 7	"	13	"	212
2 plates,	7 x 7	"	6	"	49
10 posts,	4 x 7	"	18	"	420
44 1st floor joists,	2 x 7	"	13	"	667
22 1st floor joists,	2 x 7	"	6	"	154
48 2d. "	2 x 7	"	13	"	728
24 2d. floor joists,	2 x 7	"	6	"	168
48 3d. floor joists,	2 x 7	"	13	"	728
24 3d. floor joists,	2 x 7	"	6	"	168
100 1st. floor joists,	2 x 4	"	9½	"	633
95 2d. floor joists,	2 x 4	"	8½	"	539
48 gable & ex. studs,	2 x 4	"	12	"	384
34 rafters,	2 x 7	"	22	"	872
17 collar girts,	2 x 7	"	15		298

                Total number of feet, —— 8201

    Enclosing and under floor boards, 7306 feet.
        Top floor boards, 3072 "
Shingles, 12 M. Clapboards, 3367 feet.

FULL FRAMES.

No. 164 building, 32 by 32 feet, posts 22 feet in height.

Same as No. 163, with the additional length of posts and studding. (use long plates.)

Total number of feet, 8560

Enclosing and under floor boards, 7818 feet.

Top floor boards, 3072 feet.

Shingles. 12 M. Clapboards, 4000 feet.

No. 165 building 32 x 32 feet, posts 12 feet in height.
Same as No. 163
Less 4 beams and upper floor joists.
" 6 feet in length of posts and studding, (use long plates.)

Total number of feet 6265

Enclosing and under floor boards, 5514 feet.

Top floor boards, 2048 feet.

Shingles, 12 M. Clapboards, 2400 feet.

No. 166 building, 32 by 34 feet, posts 18 feet in height

				feet.
4 sills,	7 x 7 inches,	32 feet long,		522
6 sills	7 x 7 "	13 "		318
3 short sills,	7 x 7 "	8 "		98
2 tie girts,	7 x 7 "	32 "		261
4 end girts,	7 x 7 "	16 '		261
4 side girts,	6 x 7 "	13 "		182
2 side girts,	6 x 7 "	8 "		56
4 beams	7 x 7 "	32 "		522
4 plates,	7 x 7 "	13 "		212
2 plates	7 x 7 "	8 "		65
10 posts,	4 x 7 "	18 "		420
44 1st floor joists	2 x 7 "	13 "		667
22 1st floor joists	2 x 7 "	8 "		205
48 2d. "	2 x 7 "	13 "		728
24 2d. "	2 x 7 "	8 "		224
48 3d "	2 x 7 "	13 "		728

Carried up

FULL FRAMES

24 3d floor joists,	2 x 7 inches,	8 feet long	224
104 1st. floor studs,	2 x 4 "	9½ "	658
100 2d floor studs	2 x 4 "	8½ "	566
48 gable & extra studs	2 x 4 "	12 "	384
36 rafters,	2 x 7 "	22 "	924
18 collar girts,	2 x 7 "	15 "	315

Total number of feet, —— 8540
Enclosing and under floor boards. 7688 feet.
Top floor boards, 3264 "
Shingles 12½ M. Clapboards. 3457 feet.

No. 167 Building 32 by 34 feet, posts 22 feet in height.
Same as No. 166 with the additional length of posts and studding.—Use long plates.
Total number of feet, 8909
Enclosing and under floor boards, 8216
Top floor boards, 3264
Shingles, 12½ M.   Clapboards, 4117

No. 168 Building 31 by 34 feet. posts 12 feet in height.
Same as No.166
Less 4 beams and upper joists.
" 6 feet in length of posts and studding.
(Use long plates.)   Total number of feet, 6534
Enclosing and under floor boards. 5798 feet,
Top floor boards. 2176 feet,
Shingles, 12½ M.   Clapboards, 2475 feet,

FULL FRAMES.

No. 169 building, 32 by 36 feet, posts 18 feet in height.

				Feet.
4 sills	7 x 7 inches,	32 feet long,		522
6 sills,	7 x 7 "	14 "		343
3 short sills,	7 x 7 "	8 "		98
2 tie girts,	7 x 7 "	32 "		261
4 end girts	7 x 7 "	16 "		261
4 side girts	6 x 7 "	14 . "	.	196
2 side girts	6 x 7 "	8 "		56
4 beams	7 x 7 "	32 "		522
4 plates,	7 x 7 "	14 "		228
2 plates,	7 x 7 "	8 "		65
10 posts,	7 x 7 "	18 "		420
44 first floor joists,	4 x 7 "	14 "		718
22 1st floor joists,	2 x 7 "	8 "		205
48 second floor joists,	2 x 7 "	14 "		784
24 2d " "	2 x 7 "	8 "		224
48 third floor joists	2 x 7 "	14 "		784
24 3d floor joists	2 x 7 "	8 "		224
108 1st floor studs,	2 x 4 "	9½ "		684
104 2d " studs,	2 x 4 "	8½ "		588
48 gable & extra studs	2 x 4 "	12 "		384
38 rafters,	2 x 7 "	22 "		975
19 collar girts,	2 x 7 "	15 "		332
	Total number of feet,		——	8874

Enclosing and under floor boards, 8010 feet.
Top floor boards, 3456 feet.
Shingles, 13¼ M. Clapboards, 3547 feet.

No. 170 Building, 32 by 36 feet, posts 22 feet in height.
Same as No. 169 with the additional length of posts and studding, use long plates,
Total number of feet 9255
Enclosing and under floor boards 8554 feet
Top floor boards 3456 feet.
Shingles 13¼ M. Clapboards 4227 feet.

FULL FRAMES

No. 169 Building 32 by 36 feet, posts 12 feet in height. Same as No. 170.
Less 4 Beams and upper floor joists.
" 6 feet in length of posts and studding, (use long plates.)
Total number of feet, 6796
Enclosing and under floor boards 6042 feet
Top floor boards 2304 "
Shingles 13¼ M.   Clapboards 2527 feet.

No. 172 building, 32 by 38 feet, posts 18 feet in height.

					feet.
4 sills,	7 x 7	inches,	32	feet long,	522
6 "	7 x 7	"	15	"	367
3 short sills,	7 x 7	"	8	"	98
2 tie girts,	7 x 7	"	32	"	261
4 end girts,	7 x 7	"	16	"	261
4 side girts,	6 x 7	"	15	"	210
2 side girts,	6 x 7	"	8	"	56
4 beams,	7 x 7	"	32	"	522
4 plates,	7 x 7	"	15	"	245
2 plates,	7 x 7	"	8	"	65
10 posts,	4 x 7	"	18	"	420
44 1st floor joists	2 x 7	"	15	"	770
22 1st floor joists,	2 x 7	"	8	"	205
48 2d, "	2 x 7	"	15	"	840
24 2d " "	2 x 7	"	8	"	224
48 3d " "	2 x 7	"	15	"	840
24 3d floor joists	2 x 7	"	8	"	224
110 1st floor studs	2 x 4	"	9½	"	696
106 2d floor studs	2 x 4	"	8½	"	600
48 gable & extra studs	2 x 4	"	12	"	384
40 rafters,	2 x 7	"	22	"	1026
20 collar girts,	2 x 7	"	15	"	350

Total number of feet. ——9186
Enclosing and under floor boards, 8362 feet,
Top floor boards 3648 feet.
Shingles, 13¾ M.   Clapboards, 3637 feet.

FULL FRAMES.

No 173 building, 32 by 38 feet, posts 22 feet in height.

Same as No. 172, with the additional length of posts and studding, (use long plates.)
Total number of feet, 9571
Enclosing and under floor boards, 8922 feet.
Top floor boards, 3648 feet.
Shingles, 13⅞ M. Clapboards, 4337 feet.

No. 174 building, 32 by 38 feet, posts 12 feet in height.

Same as No. 172
Less 4 beams and upper floor joists.
" 6 feet in length of posts and studding.
(Use long plates.)
Total number of feet. 7044
Enclosing and under floor boards, 6306 feet.
Top floor boards, 2432 feet.
Shingles, 13⅛ M. Clapboards, 2587 feet.

FULL FRAMES.

No. 175 building, 34 by 34 feet, posts 18 feet in height.

				Feet.
4 sills,	7 x 7 inches,	34 feet long,		556
6 sills,	7 x 7 "	14 "		343
3 short sills,	7 x 7 "	6 "		73
2 tie girts,	7 x 7 "	34 "		278
4 end girts,	7 x 7 "	17 "		278
4 side girts	6 x 7 "	14 "		196
2 side girts,	6 x 7 "	6 "		42
4 beams,	7 x 7 "	34 "		556
4 plates,	7 x 7 "	14 "		228
2 plates,	7 x 7 "	6 "		49
10 posts,	4 x 7 "	18 "		420
46 1st floor joists	2 x 7 "	14 "		751
23 1st floor joists	2 x 7 "	6 "		161
48 2d floor joists	2 x 7 "	14 "		784
24 2d floor joists	2 x 7 "	6 "		168
48 3d " "	2 x 7 "	14 "		784
24 3d " "	2 x 7 "	6 "		168
106 1st floor studs,	2 x 4 "	9½ "		671
102 2d " studs,	2 x 4 "	8½ "		578
50 gable & extra studs,	2 x 4 "	14 "		466
36 rafters,	2 x 7 "	23½ "		990
18 collar girts,	2 x 7 "	16 "		336

Total number of feet, ———8876

Enclosing and under floor boards, 8100 feet.
Top floor boards, 3468 feet.
Shingles, 13¾ M. Clapboards, 3622 feet.

No. 176 building, 34 by 34 feet, posts 22 feet in height.

Same as No. 175, with the additional length of posts and studding, (use long plates.)

Total number of feet, 9251

Enclosing and under floor boards, 8644 feet.
Top " 3468 feet.
Shingles, 13¾ M. Clapboards, 4300 feet.

FULL FRAMES.

No. 177 building, 34 by 34 feet, posts 12 feet in height.

Same as No. 175,
Less 4 beams and upper floor joists.
" 6 feet length of posts and studding. (use short plates.)
Total number of feet 6828
Enclosing and under floor boards, 6128 feet.
Top " 2312 feet.
Shingles, 13⅜ M. Clapboards, 2600 feet.

No. 178 building, 34 by 36 feet, posts 18 feet in height.

					Feet.
4 sills,	7 x 7 inches,	34	feet long,		556
6 sills,	7 x 7	"	14	"	343
3 short sills,	7 x 7	"	8	"	98
2 tie girts,	7 x 7	"	34	"	278
4 end girts,	7 x 7	"	17	"	278
4 side girts,	6 x 7	"	14	"	196
2 side girts,	6 x 7	"	8	"	56
4 beams	7 x 7	"	34	"	556
4 plates,	7 x 7	"	14	"	228
2 plates	7 x 7	"	8	"	65
10 posts,	4 x 7	"	18	"	420
46 1st floor joists,	2 x 7	"	14	"	751
23 1st floor joists	2 x 7	"	8	"	214
48 2d floor joists,	2 x 7	"	14	"	784
24 2d floor joists	2 x 7	"	8	"	224
48 3d floor joists,	2 x 7	"	14	"	784
24 3d floor joists	2 x 7	"	8	"	224
110 1st floor studs	2 x 4	"	9½	"	696
106 2d floor studs	2 x 4	"	8½	"	600
50 gable & extra studs,	2 x 4	"	14	"	466
38 rafters,	2 x 7	"	23½	"	1045
19 collar girts,	2 x 7	"	16	"	355

Total number of feet, ———— 9217
Enclosing and under floor boards, 8474 feet.
Top floor boards, 3672 feet.
Shingles. 14 M. Clapboards, 3712 feet.

FULL FRAMES.

No. 179 Building 34 by 36 feet. posts 22 feet in height. Same as No. 178 with the additional length of posts and studding. (Use long plates.)
Total number of feet, 9602
Enclosing and under floor boards, 9034 feet.
Top floor boards, 3672 feet.
Shingles, 14 M. Clapboards, 4412 feet.

No. 180 Building 34 by 36 feet. posts 12 feet in height. Same as No. 178
Less 4 beams and upper floor joists.
" 6 feet in length of posts and studding.
(Use long plates.) Total number of feet, 7097
Enclosing and under floor boards, 6010 feet.
Top floor boards, 2448 feet.
Shingles 14 M. Clapboards, 2662

No. 181 building, 34 by 38 feet, posts 18 feet in height.

				Feet.
4 sills,	7 x 7 inches,	34 feet long,	556
6 sills,	7 x 7 "	15 "	367
3 short sills,	7 x 7 "	8 "	98
2 tie girts.	7 x 7 "	34 "	278
4 end girts	7 x 7 "	17 "	278
4 side girts	6 x 7 "	15 "	210
2 side girts	6 x 7 "	8 "	56
4 beams	7 x 7 "	34 "	556
4 plates	7 x 7 "	15 "	246
2 plates,	7 x 7 "	8 "	65
10 posts,	4 x 7 "	18 "	420
46 1st floor joists,	2 x 7 "	15 "	805
23 1st "	2 x 7 "	8 "	214
48 2d floor joists,	2 x 7 "	15 "	840
24 2d "	2 x 7 "	8 "	224
48 3d floor joists,	2 x 7 "	15 "	840
24 3d "	2 x 7 "	8 "	224
112 1st floor studs	2 x 4 "	9½ "	709
108 2d "	2 x 4 "	8½ "	612
50 gable & ex. studs	2 x 4 "	14 "	466

FULL FRAME

40 rafters,	2 x 7 inches,	23½ feet long	1100
20 collar girts,	2 x 7 "	16 "	373

Total number of feet, ——— 9357

Enclosing and under floor boards, 8844 feet
Top " " 3876 feet,
Shingles, 14¾ M. Clapboards, 3800 feet.

No. 182 building, 34 by 38 feet, posts 22 feet in height.

Same as No. 181, with the additional length of psts and studding,—use long plates,

Total number of feet, 9928
Enclosing and under floor boards, 9420 feet.
Top floor boards, 3876 feet.
Shingles, 14¾ M. Clapboards, 4522 feet.

No. 183 building, 34 by 38 feet, posts 12 feet in height.

Same as No. 181,
Less 4 beams and upper floor joists.
" 6 feet in length of posts and studding.
Use long plates, Total 7352
Enclosing and under floor boards, 6688 feet.
Top floor boards, 2584 feet.
Shingles, 14¾ M. Clapboards, 2727 feet.

FULL FRAMES.

No. 184 Building 36 by 40 feet, posts 18 feet in height.

				Feet.
4 sills,	7 x 7 inches	34 feet long,		556
6 sills,	7 x 7 "	15	"	367
3 short sills,	7 x 7 "	10	"	122
2 tie girts,	7 x 7 "	34	"	278
4 end girts,	7 x 7 "	17	"	278
4 side girts,	6 x 7 "	15	"	210
2 side girts,	6 x 7 "	10	"	70
4 beams,	7 x 7 "	34	"	556
4 plates,	7 x 7 "	15	"	246
2 plates,	7 x 7 "	10	"	81
10 posts,	4 x 7 "	18	"	420
46 1st. floor joists.	2 x 7 "	15	"	805
23 1st. floor joists,	2 x 7 "	10	"	267
48 2d. floor joists,	2 x 7 "	15	"	840
24 2d. floor joists,	2 x 7 "	10	"	279
48 3d. floor joists,	2 x 7 "	15	"	840
24 3d. floor joists,	2 x 7 "	10	"	279
116 1st. floor studs,	2 x 4 "	9½	"	734
112 2d. " "	2 x 4 "	8½	"	612
50 gable & extra studs,	2 x 4 "	14	"	466
42 rafters,	2 x 7 "	23½	"	1155
21 collar girts,	2 x 7 "	16	"	382

Total number of feet, ——— 9843

Enclosing and under floor boards, 9214 feet,
Top floor boards, 4080 feet.
Shingles, 15½ M. Clapboards, 3892 feet.

No. 185 Building, 34 by 40 feet, posts 22 feet in height.
Same as No. 184 with the additional length of posts and studding, use long plates,

Total number of feet 10254
Enclosing and under floor boards 9806 feet
Top floor boards 4080 feet
Shingles 15½ M. Clapboards 4622 feet.

FULL FRAMES.

No. 186 Building 34 by 40 feet. posts 12 feet in height.
Same as No.184
Less 4 beams and upper joists.
" 6 feet in length of posts and studding.
(Use long plates.)   Total number of feet, 7584
Enclosing and under floor boards, 6966 feet,
Top floor boards, 2720 feet,
Shingles, 15½ M.   Clapboards, 2782 feet,

No. 187 building, 30 by 40 feet, posts 18 feet in height.

					Feet.
4 sills	7 x 7 inches,	30 feet long,	490
6 sills,	7 x 7 "	15 "	367
3 short sills,	7 x 7 "	10 "	122
2 tie girts,	7 x 7 "	30 "	245
4 end girts	7 x 7 "	15 "	245
4 side girts	6 x 7 "	15 "	210
2 side girts	6 x 7 "	10 "	70
4 beams	7 x 7 "	30 "	490
4 plates,	7 x 7 "	15 "	245
2 plates,	7 x 7 "	10 "	81
10 posts,	4 x 7 "	18 "	420
40 first floor joists,	2 x 7 "	15 "	700
20 1st floor joists,	2 x 7 "	10 "	233
44 second floor joists,	2 x 7 "	15 "	770
22 2d " "	2 x 7 "	10 "	256
44 third floor joists	2 x 7 "	15 "	770
22 3d floor joists	2 x 7 "	10 "	256
110 1st floor studs,	2 x 4 "	9½ "	696
106 2d " studs,	2 x 4 "	8½ "	600
15 gable & extra studs	2 x 4 "	12 "	360
42 rafters,	2 x 7 "	20½ "	1008
21 collar girts.	2 x 7 "	14 "	343

Total number of feet, —— 8977

Enclosing and under floor boards, 8234 feet.
Top floor boards, 3600 feet.
Shingles, 13½ M. Clapboards, 3590 feet.

FULL FRAMES

No. 193 building, 36 by 40 feet, posts 18 feet in height.

				Feet.
4 sills,	7 x 7 inches,	36 feet long,	588
6 sills,	7 x 7 "	15 "	367
3 short sills,	7 x 7 "	10 "	122
2 tie girts,	7 x 7 "	36 "	294
4 end girts	7 x 7 "	18 "	294
4 side girts	6 x 7 "	15 "	210
2 side girts	6 x 7 "	10 "	70
4 beams	7 x 7 "	36 "	588
4 plates,	7 x 7 "	15 "	245
2 plates	7 x 7 "	10 "	81
10 posts,	4 x 8 "	18 "	480
50 first floor joists,	2 x 7 "	15 "	875
25 first floor joists	2 x 7 "	10 "	291
52 second " "	2 x 7 "	15 "	910
26 second floor joists	2 x 7 "	10 "	303
52 third " "	2 x 7 "	15 "	910
26 3d. "	2 x 7 "	10 "	303
120 1st floor studs,	2 x 4 "	9½ "	760
116 2d floor studs	2 x 4 "	8½ "	656
52 gable and extra studs	2 x 4 "	14 "	475
42 rafters,	2 x 7 "	24½ "	1197
21 collar girts,	2 x 7 "	18	441

Total number of feet, ———10460

Enclosing and under floor boards, 9662 feet.
Top floor boards, 4320 feet.
Shingles, 16¼ M. Clapboards, 4045 feet.

No. 194 building, 36 by 40 feet, posts 22 feet in height.
Same as No. 193, with the additional length of posts and studding, (use long plates.)

Total number of feet, 10886
Enclosing and under floor boards, 10270 feet.
Top " 43200 feet.
Shingles, 16¼ M. Clapboards, 4800 feet.

FULL FRAMES.

114 1st floor studs,	2 x 4	"	9½	"	722
110 2d " studs,	2 x 4	"	8½	"	622
48 gable & extra studs,	2 x 4	"	12	"	384
42 rafters,	2 x 7	"	22	"	1078
21 collar girts,	2 x 7		15	"	367

Total number of feet, ———9426

Enclosing and under floor boards, 8714 feet.

Top floor boards, 3840 feet.

Shingles, 14⅝ M. Clapboards, 3727 feet.

No. 191 building, 32 by 40 feet, posts 22 feet in height.

Same as No. 190, with the additional length of posts and studding,

Use long plates.   Total number of feet 9823

Enclosing and under floor boards, 9290 feet.

             Top " 3840 feet.

Shingles 14⅝ M. Clapboards, 4447 feet.

No. 292 Building 32 by 40 feet, posts 12 feet in height.
Same as No. 190

Less 4 beans and upper floor joists.

6 feet in length of posts and studding.

(Use long plates.)   Tital number feet, 7229

Enclosing and under floor boards, 6570 feet.

                Tob " " 2560 feet.

Shingles, 14⅝ M.   Clapboards 2648 feet.

FULL FRAMES

No. 188 Building 30 by 40 feet posts 22 feet in height.
Same as No. 187 with the additional length of posts and studding. (Use long plates.)
Total number of feet, 9362
Enclosing and under floor boards, 8732 feet.
Top " " 3600 feet.
Shingles, 13½ M.   Clapboards, 4315 feet.

No. 189 Building 30 by 40 feet, posts 12 feet in height.
Same as No. 187
Less 4 beams and upper floor joists.
" 6 feet in length of posts and studding.
(Use long plates.)   Total number of feet, 6881 feet
Enclosing and under floor boards, 6194 feet.
Top " " 2400 feet.
Shingles, 13½ M.   Clapboards, 2540 feet.

No. 190 building, 32 by 40 feet, posts 18 feet in height.

				Feet.
4 sills,	7 x 7 inches,	32 feet long,		522
6 sills,	7 x 7 "	15 "		367
3 short sills,	7 x 7 "	10 "		122
2 tie girts,	7 x 7 "	32 "		261
4 end girts,	7 x 7 "	16 "		261
4 side girts	6 x 7 "	15 "		210
2 side girts,	6 x 7 "	10 "		70
4 beams,	7 x 7 "	32 "		522
4 plates,	7 x 7 "	15 "		245
2 plates,	7 x 7 "	10 "		81
10 posts,	4 x 7 "	18 "		420
44 1st floor joists	2 x 7 "	15 "		770
22 1st floor joists	2 x 7 "	10 "		256
46 2d floor joists	2 x 7 "	15 "		805
23 2d floor joists	2 x 7 "	10 "		268
46 3d " "	2 x 7 "	15 "		805
23 3d " "	2 x 7 "	10 "		268

FULL FRAMES

No. 195 Building 36 by 40 feet, posts 12 feet in height. Same as No. 193.
Less 4 Beams and upper floor joists.
" 6 feet in length of posts and studding, (use long plates.)
Total number of feet, 8043
Enclosing and under floor boards 7310 feet
Top floor boards 2880 "
Shingles 16¼ M.   Clapboards 2900 feet.

No. 196 building, 38 by 40 feet, posts 18 feet in height.

				feet
4 sills,	7 x 7 inches,	38	feet long,	620
8 "	7 x 7 "	15	"	490
4 short sills,	7 x 7 "	10	"	163
2 tie girts,	7 x 7 "	38	"	310
6 end girts,	7 x 7 "	12⅔	"	310
4 side girts,	6 x 7 "	15	"	210
2 side girts,	6 x 7 "	10	"	70
4 beams,	7 x 7 "	38	"	620
4 plates,	7 x 7 "	15	"	245
2 plates,	7 x 7 "	10	"	81
12 posts,	4 x 8 "	18	"	576
52 1st floor joists	2 x 7 "	15	"	910
26 1st floor joists.	2 x 7 "	10	"	302
54 2d, "	2 x 7 "	15	"	945
27 2d " "	2 x 7 "	10	"	314
54 3d " "	2 x 7 "	15	"	945
27 3d floor joists	2 x 7 "	10	"	314
124 1st floor studs	2 x 4 "	9½	"	785
120 2d floor studs	2 x 4 "	8½	"	680
56 gable & extra studs	2 x 4 "	14	"	504
42 rafters,	2 x 7 "	25½	"	1248
21 collar girts,	2 x 7 "	18	"	441

Total number of feet. —— 11083

Enclosing and under floor boards, 10110 feet,
Top floor boards 4560 feet.
Shingles, 16⅞ M. Clapboards, 4197 feet.

FULL FRAMES.

No. 197 building, 38 by 40 feet, posts 22 feet in height.

Same as No. 196, with the additional length of posts and studding, (use long plates.)
       Total number of feet, 11525
Enclosing and under floor boards, 10734 feet.
     Top floor boards, 4560 feet.
Shingles, 16⅞ M. Clapboards, 4977 feet.

No. 198 building, 38 by 40 feet, posts 12 feet in height.

Same as No. 196
 Less 4 beams and upper floor joists.
  " 6 feet in length of posts and studding.
(Use long plates.)
      Total number of feet, 8443
    Enclosing and under floor boards, 7654 feet.
      Top floor boards, 3060 feet.
Shingles, 16⅞ M. Clapboards. 3027 feet.

FULL FRAMES

No. 199 building 40 by 40 feet, posts 18 feet in height.

				Feet.
2 cross sills,	8 x 8 inches,	40 feet long,	426
2 end "	7 x 7 "	40 "	326
8 "	7 x 7 "	15 "	490
4 short sills,	7 x 7 "	10 "	163
2 tie girts,	7 x 7 "	40 "	373
6 end girts,	7 x 7 "	13$\frac{1}{3}$ "	330
4 side girts,	6 x 7 "	15 "	210
2 side girts,	6 x 7 "	10 "	70
4 beams,	7 x 7 "	40 "	653
4 plates,	7 x 7 "	15 "	245
2 plates.	7 x 7 "	10 "	81
12 posts,	4 x 8 "	18 "	576
54 1st floor joists,	2 x 8 "	15 "	1080
27 1st floor joists,	2 x 8 "	10 "	360
56 2d. "	2 x 7 "	15 "	980
28 2d. floor joists,	2 x 7 "	10 "	326
56 3d. floor joists,	2 x 7 "	15 "	980
28 3d. floor joists,	2 x 7 "	10 "	326
125 1st, floor studs,	2 x 4 "	9$\frac{1}{2}$ "	779
125 2d. floor studs,	2 x 4 "	8$\frac{1}{2}$ "	709
58 gable & ex. studs,	2 x 4 "	14 "	522
42 rafters,	2$\frac{1}{2}$ x 7 "	27 "	1652
21 collar girts,	2 x 7 "	18 "	441

Total number of feet, ——— 12098

Enclosing and under floor boards, 10602 feet.

Top floor boards, 4800   "

Shingles. 17$\frac{7}{8}$ M. Clapboards, 4350 feet.

No. 200 building, 40 by 40 feet, posts 22 feet in height.

Same as No. 199, with the additional length of posts and studding,—use long plates.

Total number of feet, 12543

Enclosing and under floor boards, 11242 feet.

Top floor boards, 4800 feet.

Shingles, 17$\frac{7}{8}$ M.   Clapboards, 5140 feet.

FULL FRAMES.

No. 201 building 40 x 40 feet, posts 12 feet in height.
  Same as No. 199
Less 4 beams and upper floor joists.
  " 6 feet in length of posts and studding, (use long plates.)
                              Total number of feet 9336
Enclosing and under floor boards, 8402 feet.
          Top floor boards, 3200 feet.
Shingles, 17⅞ M. Clapboards, 3150 feet.

No. 202 building. 40 by 50 feet, posts 18 feet in height

					feet.
5 sills,	8 x 8 inches,	40 feet long,	1066
16 sills	8 x 8	"	12½	"	1066
3 tie girts,	7 x 8	"	40	"	560
6 end girts,	7 x 8	"	13½	"	378
8 side girts,	7 x 8	"	12½	"	466
5 beams	7 x 8	"	40	"	933
8 plates,	7 x 7	"	12½	"	408
14 posts,	5 x 9	"	18	"	945
108 1st floor joists	3 x 8	"	12½	"	2700
116 2d. "	2½ x 8	"	12½	"	2407
116 3d "	2 x 8	"	12½	"	1933
140 1st. floor studs,	2 x 5	"	9½	"	1109
140 2d floor studs	2 x 5	"	8½	"	991
58 gable & extra studs	2 x 5	"	14	"	676
52 rafters,	2½ x 7	"	27	"	2045
26 collar girts,	2 x 7	"	20	"	606

                  Total number of feet,   ——— 18289
Enclosing and under floor boards. 12702 feet.
          Top floor boards, 6000 "
Shingles 22. M. Clapboards. 4800 feet.

FULL FRAMES.

No. 203 Building, 40 by 50 feet, posts 22 feet in height.

Same as No. 202 with the additional length of posts and studding, use long plates,

Total number of feet 18965
Enclosing and under floor boards 13422 feet
Top floor boards 6000 feet
Shingles 22 M. Clapboards 5700 feet.

No. 204 Building 40 by 60 feet, posts 18 feet in height.

					Feet.
6 sills,	8 x 8 inches	40 feet long,	1279
20 short sills,	8 x 8 "	12 "	1280
4 tie girts,	7 x 8 "	40 "	746
6 end girts,	7 x 8 "	13½ "	378
10 side girts,	7 x 8 "	12 "	560
5 beams,	7 x 8 "	40 "	1120
10 plates,	7 x 7 "	12 "	490
16 posts,	5 x 9 "	18 "	1066
135 1st. floor joists.	3 x 8 "	12 "	3240
145 2d. floor joists,	2½ x 8 "	12 "	2900
145 3d. floor joists,	2 x 8 "	12 "	2320
160 1st. floor studs,	2 x 5 "	9½ "	1013
160 2d. " "	2 x 5 "	8½ "	906
58 gable & extra studs,	2 x 5 "	14 "	676
62 rafters,	2½ x 7 "	27 "	2418
31 collar girts,	2 x 7 "	20 "	723

Total number of feet, ——— 21115
Enclosing and under floor boards, 14816 feet,
Top floor boards, 7200 feet.
Shingles, 26¼ M. Clapboards, 5250 feet.

FULL FRAMES

No. 205 Building 40 by 60 feet posts 22 feet in height. Same as No. 204 with the additional length of posts and studding. (Use long plates.)

Total number of feet, 21888

Enclosing and under floor boards, 15616 feet.
Top " " 7200 feet.
Shingles, 26¼ M.   Clapboards, 6250 feet.

No. 206 building, 45 by 60 feet, posts 18 feet in height.

				Feet
6 sills,	8 x 8 inches,	45 feet long,		1440
20 short sills,	8 x 8 "	12 "		1280
4 tie girts,	7 x 9 "	45 "		945
6 end girts,	7 x 8 "	15 "		420
10 side girts,	7 x 8 "	12 "		560
6 beams	7 x 8 "	45 "		1440
10 plates,	7 x 7 "	12 "		490
16 posts.	5 x 9 "	18 "		1066
150 1st floor joists,	3 x 8 "	12 "		3600
155 2d floor joists,	2½ x 8 "	12 "		3100
10 2d tie joists	4 x 8 "	12 "		320
150 3d floor joists	2 x 8 "	12 "		2400
10 3d " tie joists,	4 x 8 "	12 "		320
165 1st floor studs	2 x 5 "	9½ "		1045
165 2d floor studs	2 x 5 "	8½ "		935
65 gable & extra studs,	2 x 5	16 "		867
62 rafters.	2¼ x 8 "	30 "		3100
31 collar girts,	2 x 7 "	22 "		795

Total number of feet, ——— 24123

Enclosing and under floor boards, 16520 feet.
Top floor boards, 8100 feet.
Shingles. 29½ M. Clapboards, 5725 feet.

BARN FRAME WITH GIRTS.

No. 207 Building 45 by 60 feet. posts 22 feet in height.
Same as No. 206, with the additional length of posts and studding.
(Use long plates.)   Total number of feet, 24918
Enclosing and under floor boards, 17360 feet,
Top floor boards, 8100 feet,
Shingles, 29½ M.   Clapboards, 6775 feet,

No. 208 barn 26 by 30 feet, posts 16 feet in height.

						Feet.	
4 sills	8 x 8 inches,	26 feet long,	554	
6 side sills,	8 x 8	"	10	"	320
3 tie sills,	7 x 7	"	10	"	122
12 posts,	7 x 7	"	16	"	784
8 end & tie girts,	7 x 7	"	13	"	424
6 side girts	6 x 7	"	10	"	210
6 plates,	7 x 7	"	10	"	244
4 beams	7 x 7	"	26	"	424
12 small girts	4 x 4	"	10	"	160
10 small girts,	4 x 4	"	13	"	175
12 braces	4 x 4	"	5	"	80
48 first floor joists,	3 x 8	"	10	"	960
45 second floor joists,	3 x 6	"	10	"	675
6 2d " tie	4 x 6	"	10	"	120
32 rafters.	2 x 7	"	17½	"	656

Total number of feet, —— 5908

Enclosing boards for body & gables 2075 feet.
Boards for roof 1100 feet.
Lining floor boards, 1560 feet
Top floor " 1560 feet surface.
—— 6295
Battings, 5 in. wide about 1000 feet.
Shingles, 8⅞ M.

BARN FRAMES WITH GIRTS

No. 208, if studded and clapboarded.

110 studs, 2 x 4 inches 16 feet long	1173
Taking out the 4 x 4 small girts	335
Total number of feet	838

which gives 1838 feet more faame lumber, if studded.
Clapboards, 2480 feet.

No. 209 barn 30 by 36 feet, posts 16 feet in height.

					Feet.
4 sills,	8 x 8 inches,	30 feet long,			640
6 side sills,	8 x 8 "	12	"		384
3 tie sills,	7 x 8 "	12	"		168
12 posts,	7 x 7 "	16	"		784
8 end & tie girts,	7 x 7 "	15	"		490
6 side girts	6 x 7 "	12	"		252
12 small girts,	4 x 4 "	12	"		192
10 small girts	4 x 4 "	15	"		200
12 braces	4 x 4 "	5	"		80
6 plates,	7 x 7 "	12	"		294
4 beams,	7 x 7 "	30	"		490
57 1st. floor joists,	3 x 8 "	12	"		1368
54 2d. floor joists,	3 x 6 "	12	"		972
6 2d. floor tie "	4 x 6 "	12	"		144
38 rafters,	2 x 7 "	20¼	"		912

Total number of feet, 7370

Enclosing boards for body & gables,   2475 feet.
Roof boards,   1575 feet.
Lining boards,   2160 feet.
Top floor boards, 1575 feet. surface
Battings, 5 inches wide about 1050 feet.
Shingles, 12¼ M-

BARN FRAMES WITH GIRTS.

No. 209, if studded and clapboarded.
130 studs, 2 x 4 inches, 16 feet long, 1386
Less the 4 by 4 small girts, 392
───
994

which gives 994 feet more frame lumber, if studded. Clapboards, 3060 feet.

No. 210 barn, 30 by 40 feet, posts 16 feet in height.

					Feet.
4 sills,	8 x 8 inches,	30 feet long,			640
6 sills,	8 x 8 "	14 "			448
3 sills,	8 x 8 "	12 "			192
12 posts,	7 x 7 "	16 "			784
4 end & tie girts	7 x 7 "	15 "			490
4 side girts	7 x 7 "	14 "			228
2 side girts	7 x 7 "	12 "			98
8 small girts,	4 x 4 "	14 "			149
4 small girts,	4 x 4 "	12 "			64
12 small girts,	4 x 4 "	15 "			240
28 braces,	4 x 4 "	5 "			186
4 plates,	7 x 7 "	14 "			228
2 plates	7 x 7 "	12 "			98
4 beams	7 x 7 "	30 "			490
38 first floor joists,	3 x 8 "	14 "			1064
19 first floor joists	3 x 8 "	12 "			456
40 second "	" 3 x 6 "	14 "			840
20 second floor joists	3 x 6 "	12 "			360
42 rafters,	2 x 6 "	20½ "			861
Lumber for perline plates & posts, about					500
	Total number of feet,			───	8422

Enclosing boards for body and gables, 2600 feet.
Roof boards, 1762 feet.
Lining floor boards, 2400 feet.
Top flooring, 2400 feet, surface.
Battings, 5 in. wide, about 1250 feet. Shingles, 13¼ M.

BARN FRAME WITH GIRTS.

No 210 If studded and clapboarded.
135 studs 2 x 4 inches 16 feet long   1440 feet.
Less the 4 x 4 small girts,   453 feet

Which leaves 987 feet more, frame lumber if studded   987 feet.
Clapboaads, 3225 feet.

No. 211 Barn, 36 by 40 feet, posts 16 feet in height.

					Feet.
4 sills,	8 x 8	inches,	36 feet	long,	768
8 sills,	8 x 8	"	14	"	597
4 sills,	8 x 8	"	12	"	256
16 posts,	7 x 7	"	16	"	1045
12 end & mid girts.	7 x 7	"	12	"	588
4 side girts	7 x 7	"	14	"	228
2 side girts	7 x 7	"	12	"	98
8 small girts	4 x 4	"	14	"	149
20 small girts,	4 x 4	"	12	"	320
2 plates,	7 x 7	"	12	"	98
4 plates.	7 x 7	"	14	"	228
32 braces,	4 x 4	"	5	"	213
4 beams	7 x 7	"	36	"	568
46 1st floor joists,	3 x 8	"	14	"	1288
23 1st "	3 x 8	"	12	"	522
46 2d floor joists,	3 x 6	"	14	"	966
23 2d floor joists,	3 x 6	"	12	"	414
4 tie joists,	4 x 6	"	14	"	112
42 rafters.	2 x 6	"	24½	"	1029
Lumber for perline posts & plates			about		540

Total number of feet,   10007

Enclosing boards for body and gables,   2932 feet,
Roof boards,   2100 feet,
Lining floor boards,   2880 feet,
Top flooring,   Surface,   2880 feet,
Battings, 5 inches wide.   about   1300 feet,
Shingles, 16¼ M.

BARN FRAME WITH GIRTS.

No, 211, if studded and clapboarded.
146 studs, 2 x 4 inches, 16 feet long.   1557 feet.
    Less the 4 x 4 small girts.   469 feet.
                                                    1088

Which gives 1088 feet more frame lumber, if studded.
Clapboards, 3650 feet.

No. 212, barn 40 by 40 feet, posts 16 feet in length.

				Feet
4 sills,	8 x 10	"	40 "	1066
8 sill,	8 x 8	"	14 "	597
4 sills,	8 x 8	"	12 "	256
16 posts,	7 x 7 -	"	16 "	1045
12 end and tie girts,	7 x 7	"	13½ "	649
4 side girts,	7 x 7	"	14 "	228
2 side girts,	7 x 7	"	12 "	98
8 small girts,	4 x 4	"	14 "	149
4 small girts,	4 x 4	"	12 "	64
16 small girts,	4 x 4	"	13½ "	284
32 braces,	4 x 4	"	5 "	213
4 plates,	7 x 7	"	14 "	228
2 plates,	7 x 7	"	12 "	98
4 beams,	7 x 7	"	40 "	653
54 1st. floor joists,	3 x 8	"	14 "	1512
27 1st. floor joists,	3 x 8	"	12 "	648
54 2d floor joists,	3 x 6	"	14 ".	1134
4 " tie "	4 x 6	"	14 "	112
2 " "	4 x 6	"	12 "	48
27 2d. floor joists,	3 x 6	"	12 "	486
42 rafters,	3 x 6	"	27 "	1701

Lumber for perline posts and plates, about   550
         Total number of feet ——— 11819
    Enclosing boards for body & gables, 3210 feet
      Roof and lining floor boards, 5522 feet.
        Top floors, surface, 3200 feet
      Batting, 5 inches wide, about 1400 feet.

BARN FRAMES WITH GIRTS

No. 212, if studded and clapboarded.

155 studs, 2 x 4 inches 16 feet long	1653
Taking out the 4 x 4 small girts	497
Total number of feet	1156

which gives 1156 feet more frame lumber, if studded. Clapboards, 3950 feet.

No. 213 barn 40 by 50 feet, posts 16 feet in height.

				Feet.
5 sills,	8 x 10 inches,	40 feet long,		1333
16 sills,	8 x 8 "	12½ "		1066
20 posts,	7 x 7 "	16 "		1407
15 end & tie girts,	7 x 8 "	13⅓ "		932
8 side girts	7 x 8 "	12½ "		466
16 small girts,	4 x 4 "	12½ "		266
16 small girts,	4 x 4 "	13⅓ "		284
38 braces,	4 x 4 "	5 "		253
8 plates,	7 x 7 "	12½ "		395
5 beams,	7 x 7 "	40 "		816
108 1st. floor joists,	3 x 8 "	12½ "		2700
108 2d. floor joists,	3 x 7 "	12½ "		2349
8 2d. floor tie "	4 x 7 "	12½ "		233
52 rafters.	3 x 6 "	27 "		2106
Lumber and perline posts and plates,			about	750

Total number of feet, 15346

Enclosing boards for body & gables,   3600 feet.
Roof and lining floor boards,   6900 feet,
   Top floor boards, surface 4000 feet.
   Battings, 5 inches wide about 1520 feet.
Shingles, 22 M.

BARN FRAMES WITH GIRTS.

No. 213, if studded and clapboarded.
170 studs, 2 x 4 in. 16 feet long,   1812 feet.
   Less the 4 x 4 small girts,   550
                                              1262

Which gives 1262 feet more frame lumber, if studded
Enclosing and under floor boards, 10516 feet.
    Shingles 22 M. Clapboards 4350 feet.

No. 214 Barn,   40 by 60 feet, posts 16 feet in height.

				Feet.
6 sills,	8 x 10 inches	40 feet long,		1600
20 short sills,	8 x 8 "	12 "		1280
24 posts,	7 x 7 "	16 "		1688
18 end & tie girts,	7 x 8 "	13½ "		1118
10 side girts,	7 x 8 "	12 "		560
20 small girts,	4 x 4 "	12 "		320
16 small girts,	4 x 4 "	13½ "		284
44 braces,	4 x 4 "	5 "		293
10 plates,	7 x 7 "	12 "		490
6 beams,	7 x 7 "	40 "		980
135 1st. floor joists,	3 x 8 "	12 "		3240
135 2d. floor joists,	3 x 7 "	12 "		2835
10 " tie "	4 x 7 "	12 "		280
62 rafters,	3 x 6 "	27 "		2511

   Lumber for perline posts and plates,   about 950
         Total number of feet, ——— 18429

Enclosing boards for body & gables, 3850 feet,
     Roof and lining boards, 8200 feet.
       Top floors, surface,   4800 feet.
        Battings, 5 in. wide, about 1650 feet.
Shingles, 26¼ M.

BARN FRAMES WITH GIRTS.

No. 214 If studded and clapboarded.
185 studs, 2 x 5 inches 16 feet long,   2466
   Less the 4 x 4 small girts,   604
                                                                   ——1862

Which gives 1862 feet more frame lumber if studded. Clapboards, 4812 feet.

No. 215 barn 40 by 70 feet, posts 16 feet in height.

				feet.
7 sills,	8 x 10 inches,	40 feet long,	1866
24 short sills,	8 x 8 "	11½ "	1452
28 posts,	7 x 7 "	16 "	1969
21 end & tie girts,	7 x 8 "	13½ "	1305
12 side girts,	7 x 8 "	11½ "	634
24 small girts,	4 x 4 "	11½ "	360
16 small girts,	4 x 4 "	13½ "	284
50 braces,	4 x 4 "	5 "	333
12 plates,	7 x 7 "	11½ "	555
7 beams,	7 x 7 "	40 "	1143
162 1st floor joists	3 x 8 "	11½ "	3672
162 2d floor joists,	3 x 7 "	11½ "	3200
12 2d, "tie"	4 x 7 "	11½ "	318
72 rafters,	3 x 6 "	27 "	2916
Lumber for perline posts and plates.		about	1150
		Total number of feet.	——21157

Enclosing and under floor boards, and gables   4150 feet.
  Roof and lining floor boards,   9542 feet.
    Top floors,   Surface,   .5600 feet.
   Batting 5 inches wide about 1800 feet.
   Shingles, 30½ M.

BARN FRAMES WITH GIRTS.

No. 215 If studded and clapboarded.
200 studs 2 x 5 inches 16 feet long,   2666
   Less the 4 x 4 small girts,   644
                                                  ——— 2022

Which gives 2022 feet more frame, lumber if studded.
Clapboards, 5187 feet.

No. 216 barn 45 by 84 feet, posts 16 feet in height

					feet.
8 sills,	8 x 10 inches,	45	feet long,		2400
28 short sills	8 x 10 "	12	"		2240
32 posts,	8 x 8 "	16	"		2720
24 end & tie girts,	8 x 8 "	15	"		1920
14 side girts,	7 x 8 "	12	"		784
28 small girts,	4 x 6 "	12	"		672
16 small girts,	4 x 6 "	15	"		450
56 braces,	4 x 4 "	5	"		374
14 plates,	7 x 8 "	12	"		784
8 beams	7 x 8 "	45	"		1680
224 1st floor joists	3 x 10 "	12	"		6720
224 2d. "	3 x 7 "	12	"		4704
14 2d " tie "	4 x 7 "	12	"		392
86 rafters,	3 x 6 "	30	"		3870

Lumber for preline posts and plates,   about 1350
                  Total number of feet,   ——— 31060
    Enclosing boards for body & gables, 4893 feet
        Roof and lining floor boards, 12840 feet.
           Top floors, surface,   7560 feet.
        Battings 5 in. wide, about 2100 feet.
Shingles 40⅔ M.

BARN FRAME WITH GIRTS.

No. 216, if studded and clapboarded,
235 studs 2 x 5 in., 16 feet long,   3133 feet.
     Less the 4 x 6 small girts, 1122
                                                 2011

Which gives 2011 feet more frame lumber if studded.
The small girts in this frame are 4 by 6 inches; in all preceding frames the small girts are 4 by 4, which accounts for there being less difference.
Clapboards, 6116 feet.

No. 217 barn 50 by 100 feet, posts 16 feet in height.

				Feet.
9 cross sills	10 x 12 inches,	50 feet long,		4500
40 side sills,	8 x 10 "	12½ "		3300
45 posts,	8 x 10 "	16 "		4800
36 end & tie girts,	8 x 10 "	12 "		2880
16 side girts,	7 x 8 "	12½ "		933
32 small girts,	4 x 6 "	12½ "		800
20 small girts,	4 x 6 "	12 "		480
64 braces,	4 x 4 "	5 "		420
16 plates,	7 x 8 "	12½ "		933
9 beams	7 x 8 "	50 "		2100
282 first floor joists,	3 x 10 "	12½ "		8812
282 second floor joists,	3 x 8 "	12½ "		7050
24 2d " tie "	4 x 8 "	12½ "		800
102 rafters,	3 x 7 "	33 "		5890
Lumber for perline posts and plates about				1800
		Total number of feet,	———	45504

Enclosing boards for body & gables 5750 feet.
Roof and lining floor boards, 16864 feet.
Top floor surface. 10000 feet
Battings, 5 in. wide about 2400 feet.
Shingles, 52¾ M.

The same if studded and clapboarded.

 270 studs, 2 x 5 inches 16 feet long   3600
 Less the 4 x 6 " small girts,   1280
                 ————
                 2420

Which gives 2420 feet more frame lumber if studed.
 Clapboards, 7187 feet.

In the preceding bills of lumber for buildings of so great a variety, it is not to be expected that the size of the lumber in in every bill will correspond with the wants of every building required to be built of such dimensions. But for houses, barns and any building required to hold an ordinary amount of weight the timber as billed (for the different size building that are billed in this book,) are of proper size and dimensions. Neither is it to be expected that the author, in all cases, has made the same allowance for waste, extra studding, &c., as to exactly agree with the ideas of every builder in the land; but the difference must be very slight.

The object of these bills to old contractors are to greatly lessen the time and labor in making an estimate, as the variety being so great, one can hardly fail of finding the desired size. For young contractors and workmen, its assistance will be of still greater value. For farmers, mechanics and any one who may wish to know the amount of frame lumber required for buildings of various dimensions, can, by referring to this book, find the desired information and in connection with other tables it is hoped the book will be of great value to all.

CONTRACTORS' AND BUILDERS' ESTIMATE.

There is but one true method by which to calculate the cost of building, work and material which is practical, safe and correct. In all the practical works of Architecture and builder's books ever published, all of which abound in valuable information, not one of them treat on or give but little, if any, information to the workman the art of calculating the cost of work and material; neither is it ever taught by the master workman, at the bench; therefore it leaves the workman entirely in the dark in regard to this important part of carpentry. Workmen, by habit, acquire the knowledge of doing work, but there are few men who can calculate the cost of work and material readily and correctly. This of necessity must be true of every man who has not had experience, as it requires many years practical experience to become familiar with all the secrets connected with this part of the trade. There are many men who have worked at the trade all their days and are good workmen, that have little idea as to the cost of erecting buildings, or the proper method of calculating the cost.

There are many contractors also who have done business for years that have no practical system by which to get at the cost of building. Such men labor under great disadvantage and inconvenience often losing money on jobs. Therefore the necessity that every man should as far as possible understand all the important points connected with the business he intends to follow, thoroughly understand all its its details that he can rely upon his own judgment. It will not answer to half know your business and rely upon others for the other half, if you wish to succeed ; this proves true in all kinds of business.

The art of calculating the cost of building being a part of carpentry all instructions being omitted, there seems to be a demand that this want should be supplied, in view of which the author endeavors to supply in part.

This part of the work is especially designed for the young contractor or workman, that he may become familiar with all the little things connected with this art, which are especially the most important. It is impossible for a man to become a successful contractor without first becoming familiar with the minor rules connected with the business.

The deficiencies of all works on carpentry is, they are only available by professional architects or by men who have had great experience; they seem to take it for granted that all men understand the business; hence the young contractor cannot find the assistance which he most needs. The author speaks from experience, having realized the difficulties every young contractor must encounter the first two or three years, in order to compete with those who have had more experience.

CLOSE ESTIMATE.

For making a close estimate, it is necessary to make a careful estimate of each and every article that is connected with the building or job to be done. Figure each item separately is the motto of the most practical builders of the country. No young or old contractor can be too careful upon this point, which is the secret of making close calculations, as by this method any man of ordinary judgement cannot make any serious mistakes. For example, in making a close estimate, first find the number of feet of frame lumber, cost per thousand feet; second find the number feet enclosing and lining floor boards, cost per thousand. Top floors, clapboards, shingles or slate. Number of window frames and cost of each; window sash, window blinds; number of door frames, or lumber in each; number of feet of lumber for outside finish; number feet of lumber for inside finish; cost of sinks, wood or iron; cost of pumps and pipe; number of chimneys, cost of each per foot; number thousand lath; number yards of plastering, cost per yard; number pounds of nails, tin and zinc; window weights or springs; locks, knobs, butts, latches and screws; number lbs. of felt paper, (if used;) stairs and railings; mantles and brackets; painting; carpenter work &c., &c.

It is a good plan to estimate the cost of Piazzas, Bay windows, Porches, Dormer windows and all such outside finish separate; there being so many different styles and finish they need special attention.

The object in separating or making so many items in making a close estimate is especially necessary for unexperienced men, as by this method it is hardly possible to make ary great mistake. There is no other sure method for young or old contractors to estimate the cost of any job large or small, high or low cost, only as by long experience you can sometimes compare one job to another; but as a general rule this is a bad plan to adopt, as there are hardly two jobs alike.

CARPENTER WORK.

There is no rule, or process of figures which can be applied to different jobs that is practical, safe and correct, as no two men or gang of men will perform a job of work at just the same cost; neither will any one man perform the same job of work for different men or, at a more unfavorable time in the year at just the same cost. Therefore figures will not give the exact cost,

Work being the most important item in estimating the cost of building, every workman or young contractor should use great care and judgment, giving this item special attention. There is but one method that new beginners can safely adopt, and that is to calculate the cost of each part separately, as in the bill of estimate for material, &c.; first calculate the time for framing the building, whatever it is, large or small, and so on separating the work and the time for each part. For example:

ESTIMATE OF WORK.

Time for framing, raising, boarding, coving and other outside finish, shingling, clapboarding, making, setting and casing window frames, making and setting door frames, setting partition studs, casing doors, putting down base boards, laying floor fitting, hanging and trimming doors, casing sink, casing pantry shelving, casing stairs, stair railings, &c.

The object of this is that workmen may become familiar with each part; every new beginner should adopt this plan until he is well posted, whatever system he may use after years of experience; but it is safe to say however, that you will not adopt a more sure method as it is hardly possible for a man of ordinary judgment to make any great blunders by this method.

WINDOW FRAMES.

The number of feet of lumber in common window frame jambs $5\tfrac{3}{4}$ inches wide, this includes jamb sills and outside casings, 5 inches wide. 18 feet. net.

WINDOW CASINGS.

Inside casings for windows. 5 inches wide, including stool surbace and box casings. 11 feet net.

DOOR FRAMES.

The number of feet of lumber for inside door frames, jambs $4\tfrac{1}{4}$ inches wide, 2 inches thick, 12 feet net.

DOOR CASINGS.

Number of feet of lumber in common inside door casings, 5 inches wide, for casing, one side,   7½ feet, net

OUTSIDE DOOR FRAMES.

Common outside door frames, 3 by 7 feet jambs, 2 inches, thick, including stool,   25 feet, net.

CASING FOR OUTSIDE DOORS.

Casings 8 inches wide and face boards under stools, about 16 feet

Inside casings for same, 6 in. wide,   9 feet

FRONT STAIRS.

The number of feet of lumber in common flight front stairs, 9 feet studding, 8 in. risers, 10 in treads, 1⅛ in. thick, including face and apron boards,   about 110 feet, net.

BACK STAIRS.

Including face boards with treads and risers, about 86 feet.

ATTIC STAIRS.

Treads, risers and face boards,   about 76 feet,

CELLAR STAIRS.

About 50 feet.

In either flight of stairs the stringers are not included.

PANTRY.

Small pantry, 4 by 6, shelves 12 in. wide, one shelf 18 inches wide, cleats, &c.,   107 feet.

MOP, OR BASE BOARDS.

For common houses with ordinary size rooms, reckon 48 feet in length for each room, count the front hall 2 rooms, that is, if a house contains 10 finished rooms, call the hall 2 rooms, making 12 rooms; say nothing about the closets, as without too much waste there will be enough left for closets, allowing

Base boards to average 8 in. wide, gives 32 feet to a room. Net.
" " " 9 " " 36 " " "
" " " 10 " " 40 " " "

DISH, OR CHINA CLOSETS.

Of ordinary size, shelves 15 in. wide, $3\frac{1}{2}$ or 4 feet long, cleats, 25 to 30 feet, Net.

WINDOW FRAMES, (CASED.)

Window frames and casing for both sides, including stool surbase, box casing, &c., complete; casings 5 in. wide; 29 feet net.

DOOR FRAMES. (CASED.)

Inside door frames and casings, for both sides, jambs 2 in. thick, $4\frac{1}{4}$ in. wide, casings 5 in. wide,   27 feet, net.

DOOR FRAMES. (CASED.)

Outside door frames, cased both sides, plain jambs 2 in. thick, $5\frac{3}{4}$ in. wide, stool, or threshold, 10 in. wide, outside casing, 8 in. wide, inside casing 6 in. wide.   about 50 feet

JET COVING.

Wide square jet, plancher 2 feet wide, freize 2 feet wide, cornice fascia, 6 in. wide, $4\frac{1}{2}$ foot, 1 foot in length, that is, in a jet 100 feet long we get 450 feet in plancher, freize, and cornice fascia in width as mentioned above. This does not include the cornice and bed mouldings, as these mouldings cost more per foot and are generally bought separate all ready for use.

22-INCH JET.

Square jet, 22 inches wide, plancher 22 inches wide.
Frieze 22 " cornice fascia 6 inches wide, 4 1-6 feet, one foot in length

20-INCH JET.

Square jet plancher, 20 in. wide, freize 20 in. wide.
Cornice fascia, 5 in. wide, 3¾ feet, one foot in length.

18-INCH JET.

Square jet, plancher, 18 in. wide, freize 18 in. wide, cornice fascia 5 in. wide, about 3½ feet, one foot in length.

16-INCH JET.

Square jet plancher, 16 in. wide freize 16 in. wide, cornice 5 in. wide about 3 1-6 feet, one foot in length.

14-INCH JET.

Square jet plancher 14 in. wide, freize 14 in. wide, cornice fascia 5 in. wide, 2¾ feet, one foot in length.

12-INCH JET.

Square jet plancher 12 in. wide, freize 12 inch wide, cornice fascia 5 in. wide, 2 5-12 feet, one foot in length.

10-INCH JET.

Square jet plancher 10 in. wide, freize 10 in. wide, cornice fascia 4 in wide, 2 feet, one foot in length.

8-INCH JET.

Square jet plancher 8 in. wide, frieze 8 in. wide, cornice fascia 4 in. wide, 1⅔ feet one foot in length.

You will notice in all cases the plancher and frieze are of the same width; this of course is not always the case in building; this jet table is designed more particularly to show the shortest method of calculating the number of feet of lumber in any jet or coving, of whatever width it may be, by putting the plancher frieze and cornice fascia together, no matter what each of their width may be it is a much easier and shorter method. If in a jet the plancer is 20 in. wide, the frieze 24 in. wide, the cornice fascia is 5 in. wide, 4 1-12 feet, one foot in length. In case you wish to use cheap spring cornice and spring the same yourself, bed mould also add 6 in. for cornice and 3 in. for bed moulding. For example: in jet plancia 18 in. wide, frieze 20 in. wide, cornice 6 in. wide, cornice fascia 5 in. wide, bed mould 3 in. wide add these together will give 52 in. equal to 4⅓ feet one foot in length, and so on in whatever width jet you wish to use you will find this method much shorter and quicker done.

NAILS.

For 1000 shingles allow 3½ to 5 lbs. 4d. nails; or 3 to 3½ lbs 3d. nails.

For 1000 Laths, allow about 6 lbs. 3d. fine nails.
" 1000 feet clapboards, about 18 lbs, 6d. box.
" 1000 " boarding boards, 20 " 8d. common.
" 1000 " " 25 " 10d, "
" 1000 " Top floors, sq. edge 38 " 10d. floor.
" 1000 " " " 41 " 12d "
" 1000 " " matched blind nailed 35 " 10d "
" 1000 " " " " 42 " 12d "
" 10 " partition studs or studding, 1 " 10d common
" 1000 " furring 1 x 3 45 " 10d "

"	1000	furring	1 x 2	65	" 10d	"
"	1000	" pine finish	about	30	" 8d finish.	

20d nails,	$3\frac{5}{8}$ inches long	36 nails to a pound.
30d "	4	"	24	"	"
40d "	4 1-2	"	18	"	"
50d "	$5\frac{1}{4}$	"	13	"	"
60d "	6	"	9	"	"
70d "	7	"	6	"	"

Nails from different manufacturers vary somewhat in size but not to interfere with the use of the table.

SHEATHING AND FLOORING.

In estimating the number feet of matched lumber, all sheathing and matched flooring of ordinary width, add 25 per cent after getting the sq. feet you wish to cover; if sheathing is very narrow, say 2 or 2 1-2in. wide add 40 per cent. as all matched lumber is measured before it is sawed and matched.

BOARDING BOARDS,

All square edge boards for enclosing buildings, lining floors, &c., calculate the number of sq. feet to be covered and make no allowance for openings. Add 10 per cent.

TOP FLOORS,

Unmatched require from 8 to 12 per cent. allowance for waste. Matched from 20 to 30 per cent; very narrow matched 2 or $2\frac{1}{2}$ in. wide, 40 per cent.

CLAPBOARDS, LATH AND SHINGLES,

One thousand feet of spruce clapboards with a large $\frac{1}{4}$ lap, will lay 750 feet; or, the better way is to get the number sq. feet to be covered, pay no regard whatever to openings, then add ¼ to the surface; for houses with ordinary number of windows and doors it is very near correct and a much shorter method.

One thousand feet of pine clapboards will lay more, as the thin edge is thicker and does not require so great a lap.

SHINGLES.

Good 18 in. shingle will average to lay 130 feet per thousand laid 5¾ in. to the weather, they will sometimes overrun, but 130 feet is a safe estimate.
16 in. shingles, 5 in. to weather will cover 115 sq. feet.
15   "   3¾ "   "   "   100   " .

LATHS.

100 laths is estimated to cover 6 sq. yards of surface.
1000 "   "   "   60 "   "

One cask of lime will make mortar to spread the surface of about 750 laths or 45 sq. yards to a cask of lime. 200 lbs. to a cask, 80 lbs. to a bushel, 2½ bushels in a cask.

SAND.

One cask of lime to 10 bushels of sand is a safe estimate. Estimated for 200 lbs. lime to a cask.

LIME FOR BRICKWORK.

A cask of lime to 1050 bricks is considered a close estimate for chimneys. For brick walls it will lay about 1400 bricks.

MORTAR FOR PLASTERING.

Requires about the same quantity of sand to a cask of lime as for brick mortar, 8 lbs. of good dry hair to be mixed with a cask of lime; a cask of lime will cover about 45 sq. yards of surface.

BRICKS.

Bricks are usually estimated at 24 to a cubic foot; five courses to one foot in length.

    8 inch wall, 15 to a square foot.
    12 "   23 "   "
    16 "   31 "   "
    20 "   39 "   "

CHIMNEYS.

Bricks for chimneys are estimated for each foot in height, as follows:

Size of Chimney	Size of Flue	No. bricks each foot in height
16 by 16	8 by 8	30
16 by 20	8 by 12	35
16 by 24	8 by 16	40
20 by 20	12 by 12	40
20 by 24	12 by 16	45

CEMENT FOR BRICKS.

One cask of cement mixed with two casks of sand makes good mortar and will lay about 600 bricks. For concrete floors a cask of cement mixed with two casks of gravel is estimated to cover about 12 square yards surface one inch thick. One half lime mortar is better mixed with cement for laying bricks.

SHEET LEAD.

Sheet lead, 1-8 in. thick, $7\frac{1}{2}$ lbs. per square foot.
    "   1-10 "   6 "   "
    "   1-16 "   4 "   "

LEAD PIPE

Estimate lead pipe for each foot in length as follows:—
¾ inch bore, one foot in length, 1 lb. 12 oz.

1	"	"	"	2 "	5 "
1¼	"	"	"	2 "	12 "
1½	"	"	"	4 "	5 "
2	"	"	"	5 "	10 "

PLASTERER'S ALLOWANCE.

Allowances made by plasterers for doors and windows.

For doors and windows, one-half the surface of each is to be allowed.

FURRING.

A short method to estimate the number of feet of furring that may be required to furrow any given number of feet, the same being placed 16 inches from centers. For 1 x 3 furring it takes 1-5 of the number of surface feet.

In a ceiling 24 feet square it takes 115 feet or 38 furring 12 feet long, 3 feet in each, 3 times 38 equal to 114 feet. For 1 x 2 furring divide by 7 will be a close estimate. If the partitions are set before the building is furrowed, it will take a few more feet; always add 10 or 15 per cent. for waste.

PARTITION STUDS AND WALL STUDDING.

The shortest method to ascertain the number of studs required to stud any building, or any number of feet studding being placed 16 in. from centre to centre; first find the number of feet in length to be studded and divide by three-fourths, that is it takes three-fourths the number of studding there is number of feet adding one to start with; use the same rule for furring and floor joists when placed 16 inches from centers.

For inside partition apply the same rule adding about 20

per cent. for corners, door headers, &c.

Studding joists and furring that are placed 18 inches from centers, two-thirds the number of studding there is number of feet, (allowances made as above.)

PIAZZAS.

Width of floor 4 feet. In a narrow piazza of this width all rough lumber necessary, including sills, joists, rafters, roof boards, &c., $20\frac{1}{2}$ ft. in length. Finished lumber for same including flooring, matched sheathing for over head, pine finish for casing, jets &c. $14\frac{1}{4}$ feet, one foot in length, this includes everything except cornice, bed mould and lumber for columns (or posts.)

PIAZZAS.

5 foot floor all rough lumber, $23\frac{1}{2}$ feet, one foot in length. Finished lumber including floors sheathing for over head, pine finish for jetting &c., $16\frac{2}{3}$ feet, one foot in length- This does not include any mouldings or lumber for post.

In all cases these Piazza roofs are to be tinned and in no case the mouldings and lumber for the posts are included.

PIAZZAS.

6 feet floor, all rough lumber necessary for sills, joists, frame rafters, roof boards, &c.

$27\frac{2}{3}$ feet, one foot in length. Finish lumber, including floor sheathing for overhead, pine finish for casing jet, &c., $19\frac{1}{2}$ feet, one foot in length, no moulding or lumber for posts are included

PIAZZAS.

7 feet floor, all rough lumber 33 feet one foot in length: finish lumber including floors, sheathing, pine finish for casing jet,

&c., 22¼ feet, one foot in length, no mouldings or lumber for posts are included.

PIAZZAS.

8 feet floor, all rough lumber 40¾ feet one foot in length; finish lumber including floor sheathing, pine finish for casing jet, &c., 26½ feet one feet in length. This includes no moulding or lumber for posts.

BAY WINDOWS.

In an ordinary size bay window, one story high, plain cased and panelled, sills, joists, studding and all rough lumber, including rough boards, about 350 feet.

Pine finish complete including lumber for window frames about 215 feet.

This does not include the mouldings. The cost of a plain bay window with tin roof, carpenter labor at $2 00 per day, is from 50 to 55 dollars.

PORTICOS.

Porches vary so much in size and style of finish &c., that it is difficult to give much information.

TABLE OF JOISTS AND STUDDING.

LENGTH IN FEET

Inches	6 ft. ft. in.	7 ft. ft. in.	8 ft. ft. in.	10 ft. ft. in.	12 ft. ft. in.	14 ft. ft. in.	16 ft. ft. in.	18 ft. ft. in.	20 ft. ft. in.	22 ft. ft. in.	24 ft. ft. in.
2 x 2	2 0	2 4	2 8	3 4	4 0	4 8	5 4	6 0	6 8	7 4	8 0
2 x 3	3 0	3 0	4 0	5 0	6 0	7 0	8 0	9 0	10 0	11 0	12 0
2 x 4	4 0	4 8	5 4	6 8	8 0	9 4	10 8	12 0	13 4	14 8	16 0
2 x 5	5 0	5 10	6 8	8 4	10 0	11 8	13 4	15 0	16 8	18 4	20 0
2 x 6	6 0	7 0	8 0	10 0	12 0	14 0	16 0	18 0	20 0	22 0	24 0
2 x 7	7 0	8 2	9 4	11 8	14 0	16 4	18 8	21 0	23 4	25 8	28 0
2 x 8	8 0	9 4	10 8	13 4	16 0	18 8	21 4	24 0	26 8	29 4	32 0
2 x 9	9 0	10 6	12 0	15 0	18 0	21 0	24 0	27 0	30 0	33 0	36 0
2 x 10	10 0	11 8	13 4	16 8	20 0	23 4	26 8	30 0	33 4	36 8	40 0
2 x 11	11 0	12 10	14 8	18 4	22 0	25 8	29 4	33 0	36 8	40 4	44 0
2 x 12	12 0	14 0	16 0	20 0	24 0	28 0	32 0	36 0	40 0	44 0	48 0
2 x 13	13 0	15 2	17 4	21 8	26 0	30 4	34 8	39 0	43 4	47 8	52 0
2 x 14	14 0	16 4	18 8	23 4	28 0	32 8	37 4	42 0	46 8	51 4	56 0
2 x 15	15 0	17 6	20 0	25 0	30 0	35 0	40 0	45 0	50 0	55 0	60 0
2 x 16	16 0	18 8	21 4	26 8	32 0	37 4	42 8	48 0	53 4	58 8	64 0
2 x 17	17 0	19 10	22 8	28 4	34 0	39 8	45 4	51 0	56 8	62 4	68 0
2 x 18	18 0	21 0	24 0	30 0	36 0	42 0	48 0	54 0	60 0	66 0	72 0
2 x 19	19 0	22 2	25 4	31 8	38 0	44 4	50 8	57 0	63 4	69 8	76 0
2 x 20	20 0	23 4	26 8	33 4	40 0	46 8	53 4	60 0	66 8	73 4	80 0
2 x 21	21 0	24 6	28 0	35 0	42 0	49 0	56 0	63 0	70 0	77 0	84 0
2 x 22	22 0	25 8	29 4	36 8	44 0	51 4	58 8	66 0	73 4	80 8	88 0
2 x 24	24 0	28 0	32 0	40 0	48 0	56 0	64 0	72 0	80 0	88 0	96 0

JOISTS 3 BY 3 TO 3 BY 24 INCHES.

LENGTH IN FEET.

Inches	6 ft.		7 ft.		8 ft.		10 ft.		12 ft.		14 ft.		16 ft.		18 ft.		20 ft.		22 ft.		24 ft.	
	ft.	in.	ft.	in.	ft.	in.	ft.	in.	ft.	in.	ft.	in.	ft.	in.	ft.	in.	ft.	in.	ft.	in.	ft.	in.
3 x 3	4	6	5	3	6	0	7	6	9	0	10	6	12	0	13	6	15	0	16	6	18	0
3 x 4	6	0	7	0	8	0	10	0	12	0	14	0	16	0	18	0	20	0	22	0	24	0
3 x 5	7	6	8	9	10	0	12	6	15	0	17	6	20	0	22	6	25	0	27	6	30	0
3 x 6	9	0	10	6	12	0	15	0	18	0	21	0	24	0	27	0	30	0	33	0	36	0
3 x 7	10	6	12	3	14	0	17	6	21	0	24	6	28	0	31	6	35	0	38	6	42	0
3 x 8	12	0	14	0	16	0	20	0	24	0	28	0	32	0	36	0	40	0	44	0	48	0
3 x 9	13	6	15	9	18	0	22	6	27	0	31	6	36	0	40	6	45	0	49	6	54	0
3 x 10	15	0	17	6	20	0	25	0	30	0	35	0	40	0	45	0	50	0	55	0	60	0
3 x 11	16	6	19	3	22	0	27	6	33	0	38	6	44	0	49	6	55	0	60	6	66	0
3 x 12	18	0	21	0	24	0	30	0	36	0	42	0	48	0	54	0	60	0	66	0	72	0
3 x 13	19	6	22	9	26	0	32	6	39	0	45	6	52	0	58	0	65	0	71	6	78	0
3 x 14	21	0	24	6	28	0	35	0	42	0	49	0	56	0	63	0	70	0	77	0	84	0
3 x 15	22	6	26	3	30	0	37	6	45	0	52	6	60	0	67	6	75	0	82	6	90	0
3 x 16	24	0	28	0	32	0	40	0	48	0	56	0	64	0	72	0	80	0	88	0	96	0
3 x 17	25	6	29	9	34	0	42	6	51	0	59	6	68	0	76	6	85	0	93	6	102	0
3 x 18	27	0	31	6	36	0	45	0	54	0	63	0	72	0	81	0	90	0	99	0	108	0
3 x 19	28	6	33	3	38	0	47	6	57	0	66	6	76	0	85	6	95	0	104	6	114	0
3 x 20	30	0	35	0	40	0	50	0	60	0	70	0	80	0	90	0	100	0	110	0	120	0
3 x 21	31	6	36	9	42	0	52	6	63	0	73	6	84	0	94	6	105	0	115	6	126	0
3 x 22	33	0	38	6	44	0	55	0	66	0	77	0	88	0	99	0	110	0	121	0	132	0
3 x 23	34	6	40	3	46	0	57	6	69	0	80	6	92	0	103	6	115	0	126	6	138	0
3 x 24	36	0	42	0	48	0	60	0	72	0	84	0	96	0	108	0	120	0	132	0	144	0

4 BY 4 TO 4 BY 10 INCHES,

LENGTH IN FEET:

Inches.	6 ft.	7 ft.	8 ft.	10 ft.	12 ft.	14 ft.	16 ft.	18 ft.	20 ft.	22 ft.	24 ft.
	ft. in	ft. in	ft. in	ft. in	ft. in	ft. in	ft. in	ft. in	ft. in	ft. in	ft. in
4 x 4	8 0	9 4	10 8	13 4	16 0	18 8	21 4	24 0	26 8	29 4	32 0
4 x 5	10 0	11 8	13 4	16 8	20 0	23 4	26 8	30 0	33 4	36 8	40 0
4 x 6	12 0	14 0	16 0	20 0	24 0	28 0	32 0	36 0	40 0	44 0	48 0
4 x 7	14 0	16 4	18 8	23 4	28 0	32 8	37 4	42 0	46 8	51 4	56 0
4 x 8	16 0	18 8	21 4	26 8	32 0	37 4	42 8	48 0	53 4	58 8	64 0
4 x 9	18 0	21 0	24 0	30 0	36 0	42 0	48 0	54 0	60 0	66 0	72 0
4 x 10	20 0	23 4	26 8	33 4	40 0	46 8	53 4	60 0	66 8	73 4	80 0

5 BY 5 TO 5 BY 10 INCHES.

5 x 5	12 6	14 7	16 8	20 10	25 0	29 2	33 4	37 6	41 8	45 10	50 0
5 x 6	15 0	17 6	20 0	25 0	30 0	35 0	40 0	45 0	50 0	55 0	60 0
5 x 7	17 6	20 5	23 4	29 2	35 0	40 10	46 8	52 6	58 4	64 2	70 0
5 x 8	20 0	23 4	26 8	33 4	40 0	46 8	53 4	60 0	66 8	73 4	80 0
5 x 9	22 6	26 5	30 0	37 6	45 0	52 6	60 0	67 0	75 0	82 6	90 0
5 x 10	25 0	29 2	33 4	41 8	50 0	58 4	66 8	75 0	83 4	91 8	100

6 BY 6 TO 6 BY 10 INCHES,

6 x 6	18 0	21 0	24 0	30 0	36 0	42 0	48 0	54 0	60 0	66 0	72 0
6 x 7	21 0	24 6	28 0	35 0	42 0	49 0	56 0	63 0	70 0	77 0	84 0
6 x 8	24 0	28 0	32 0	40 0	48 0	56 0	64 0	72 0	80 0	88 0	96 0
6 x 9	27 0	31 6	36 0	45 0	54 0	63 0	72 0	81 0	90 0	99 0	108
6 x 10	30 0	35 0	40 0	50 0	60 0	70 0	80 0	90 0	100	110	120

CONTRACTORS' AND BUILDERS' ESTIMATE. 131

7 BY 7 TO 7 BY 12 INCHES.

LENGTH OF FEET.

Inches	6 ft. ft. in	7 ft. ft. in	8 ft. ft. in	10 ft. ft. in	12 ft. ft. in	14 ft. ft. in	16 ft. ft. in	18 ft. ft. in	20 ft. ft. in	22 ft. ft. in	24 ft. ft. in
7 x 7	24 6	28 7	32 8	40 10	49 0	57 2	65 4	73 6	81 8	89 10	98 0
7 x 8	28 0	32 8	37 4	46 8	56 0	65 4	74 8	84 0	93 4	102 8	112 0
7 x 9	31 6	36 9	42 0	52 6	63 0	73 6	84 0	94 6	105 0	115 6	126 0
7½ x 10	35 0	40 10	46 8	58 4	70 0	81 8	93 4	105 0	116 8	128 4	140 0
7 x 11	38 6	44 11	51 4	64 2	77 0	89 10	102 8	115 6	128 4	141 2	154 0
7 x 12	42 0	49 0	56 0	70 0	84 0	98 0	112 0	126 0	140 0	154 0	168 0

8 BY 8 TO 8 BY 12 INCHES.

8 x 8	32 0	37 4	42 8	53 4	64 0	74 8	85 4	96 0	106 8	117 4	128 0
8 x 9	36 0	42 0	48 0	60 0	72 0	84 0	96 0	108 0	120 0	132 0	144 0
8 x 10	40 0	46 8	53 4	66 8	80 0	93 4	106 8	120 0	133 4	146 8	160 0
8 x 11	44 0	51 4	58 8	73 4	88 0	102 8	117 4	132 0	146 8	161 4	176 0
8 x 12	48 0	56 0	64 0	80 0	96 0	112 0	128 0	144 0	160 0	176 0	192 0

9 BY 9 TO 9 BY 12 INCHES.

9 x 9	40 6	47 3	54 0	67 6	81 0	94 6	108 0	121 6	135 0	148 6	162 0
9 x 10	45 0	52 6	60 0	75 0	90 0	105 0	120 0	135 0	150 0	165 0	180
9 x 11	49 6	57 9	66 0	82 6	99 0	115 6	132 0	148 6	165 0	181 6	198 0
9 x 12	54 0	63 0	72 0	90 0	108 0	126 0	144 0	162 0	180 0	198 0	216 0

10 BY 10 TO 10 BY 12 INCHES.

10 x 10	50 0	58 4	66 8	83 4	100 0	116 8	133 4	150 0	166 8	183 4	200 0
10 x 11	55 0	64 2	73 4	91 8	110 0	128 4	146 8	165 0	183 4	201 8	220 0
10 x 12	60 0	70 0	80 0	100 0	120 0	140 0	160 0	180 0	200 0	220 0	240 0

12 BY 12 TO 12 BY 16 INCHES.

12 x 12	72 0	84 0	96 0	120 0	144 0	168 0	192 0	216 0	240 0	264 0	288 0
12 x 13	78 0	91 0	104 0	130 0	156 0	182 0	208 0	234 0	260 0	286 0	312 0
12 x 14	84 0	98 0	112 0	140 0	168 0	196 0	224 0	252 0	280 0	308 0	336 0
12 x 15	90 0	105 0	120 0	150 0	180 0	210 0	240 0	270 0	300 0	330 0	360 0
12 x 16	96 0	112 0	128 0	160 0	192 0	224 0	256 0	288 0	320 0	352 0	384 0

EXPLANATION OF THE RAFTER TABLE.

It is understood that the length of a rafter is to be measured from the two extreme points from outside of the plate to the extreme peak of roof; but in this rafter table no allowance has been made for the projection of rafters beyond the plate, or for the ridge board (if such is used.) Hence the length of a common rafter is the distance from the top and outer corner of the plate to the extreme peak of the roof. The table gives the length of any rafter in any pitch roof from 6 to 16 inches rise to one foot in any building from 8 to 50 feet in width, width of buildings running in even feet.

LENGTH OF RAFTERS IN FEET, INCHES AND ⅛ OF AN INCH.

Width of B'ld'g. in feet.	Run of rafters	6 in rise.	7 in. rise.	8 in. rise.	9 in rise.	10 in. rise.	11 in. rise.	12 in. rise.	13 in. rise.	14 in. rise.	15 in. rise.	16 in. rise.
8	4	4 5⅝	4 7½	4 9¾	5	5 2½	5 5⅛	5 7⅞	5 10¾	6 1¾	6 4⅞	6 8
10	5	5 7⅛	5 9½	6 0⅛	6 3	6 6⅛	6 9⅜	7 0⅞	7 4½	7 8¼	8	8 4
12	6	6 8½	6 11⅜	7 2½	7 6	7 9¾	8 1⅝	8 5⅞	8 10⅛	9 2⅝	9 7¼	10
14	7	7 10	8 1¼	8 6	8 9	9 1⅜	9 6	9 10¾	10 3⅞	10 9⅛	11 2½	11 8
16	8	8 11⅜	9 3⅛	9 7⅜	10	10 5	10 10¼	11 3¾	11 9½	12 3½	12 9⅝	13 4
18	9	10 0¾	10 5	10 9¾	11 3	11 8⅝	12 2½	12 8¾	13 3¼	13 10	14 4⅞	15
20	10	11 2⅛	11 7	12 0¼	12 6	13 0¼	13 6¾	14 1¾	14 9	15 4⅜	16	16 8
22	11	12 3⅝	12 8⅞	13 2⅝	13 9	14 3⅞	14 11	15 6⅝	16 2⅝	16 10⅞	17 7¼	18 4
24	12	13 5	13 10¾	14 5	15	15 7½	16 3⅜	16 11⅝	17 8¼	18 5¼	19 2½	20
26	13	14 6½	15 0⅝	15 7½	16 3	16 11	17 7⅝	18 4⅝	19 2	19 11¾	20 9¾	21 8
28	14	15 7⅞	16 2½	16 10	17 6	18 2¾	19	19 9⅝	20 7¾	21 6⅛	22 5	23 4
30	15	16 9¼	17 4⅜	18 0¼	18 9	19 6¼	20 4¼	21 2½	22 1⅜	23 0⅝	24 0⅛	25
32	16	17 10⅝	18 6¼	19 2¾	20	20 10	21 8½	22 7½	23 7	24 7	25 7⅜	26 8
34	17	19	19 8⅛	20 5⅛	21 3	22 1½	23 0¾	24 0⅝	25 0¾	26 1½	27 2½	28 4
36	18	20 1⅝	20 10	21 7⅝	22 6	23 5⅛	24 5	25 5½	26 6½	27 7⅞	28 9¾	30
38	19	21 2⅞	22	22 10	23 9	24 8¾	25 9¼	26 10½	28 0⅛	29 2⅜	30 5	31 8
40	20	22 4⅜	23 1⅞	24 0½	25	26 0⅜	27 1⅝	28 3½	29 5⅞	30 8¾	32 0¼	33 4
42	21	23 5¾	24 3¾	25 2⅞	26 3	27 4	28 5⅞	29 8⅜	30 11½	32 3¼	33 7⅜	35
44	22	24 7⅛	25 5⅝	26 5¼	27	28 7⅝	29 10⅛	31 1⅜	32 5¼	33 9⅝	35 2⅝	36 8
46	23	25 8⅝	26 7½	27 7¾	28 9	29 11¼	31 2½	32 6⅜	33 11	35 4⅛	36 9⅞	38 4
48	24	26 10	27 9⅜	28 10⅛	30	31 2⅞	32 6¾	33 11¼	35 4⅝	36 10½	38 5	40
50	25	27 11⅜	28 11¼	30 0½	31 3	32 6½	33 11	35 4¼	36 10¼	38 5	40 0¼	41 8

BRACES.

To get the length of braces by a short and easy method, where the perpendicular and horizontal runs are equal which is most always the case, multiply the run of brace by 17 and it will give you the exact length required the length to be measured from the extreme points from shoulders, or longest corners.

The length of braces as given on the square or by the rule of extracting the square root makes a brace a little too short; a small allowance should always be made. Hence to get the length of braces for regular runs, there is no other way that will give the exact length required. For example: the common run for a brace is three feet; 3 times 17 is 51 inches, the exact length; every close framer will cut the length of the brace from longest corners of shoulders; when by extracting the square root it will give for the length 50 inches and 91-100 of an inch, nearly one-eighth of an inch too short. If the runs are 5 feet; 5 times 17 is 85 inches, extract the square root and it will give for the length 84 inches and 85-100 of an inch, which is more than one-eigth of an inch too short. This rule can be safely used to obtain the length of any brace that may be required for any building.

The same rule can be applied to obtain the length of rafters that are to be on a mitre pitch, or 12 inches rise to one foot; a building 24 feet wide, one-half the width is 12 feet; 12 times 17 is 204 inches, equal to 17 feet, the length of rafter, from out side of the plate to the extreme peak of the roof, both bevels of course being alike; in rafters or braces the runs must be equal to apply this rule.

BOARD MEASURE.

Surface measure has length and breadth only, without regard to thickness.

If a board is 12 inches long and 12 inches wide it contains 144 square inches or one square foot. Rule; multiply the length by the breadth and divide the product by 12.

If a board is 16 feet long and 14 inches wide. Example:

```
        16
        14
        ——
        64
        16
        ——
12)224(18⅔ sq. feet.
        12
        ——
       104
        96
        ——
       4)8(⅔
         12
```

If a board is 19 1-2 feet long and 16 1-2 inches wide; 19 1-2 multiplied by 16 1-2 equals 322, reduced to feet by dividing by 12, 26 10-12 sq. feet

If a board is 20 feet long and 16 inches wide; 20 multiplied 16 equals 320 divided by 12 gives, reduced, 26⅔ sq. feet, or

16 inches is equal to 1⅓ feet multiplied by the length which is 20 feet; 20 feet multipled by 1⅓ feet, gives 26⅔ square feet as above.

In a floor of a room 25 feet long by 20 feet wide, 25 x 20 equals 500 feet, or, 250 x 2 equals 500 feet.

SQUARE TIMBER AND JOISTS,

Dimension lumber is measured board or inch measure.

If a plank is 12 inches wide, 4 inches thick and 16 feet long, how many feet, board measure?

```
        12
         4
        ——
        48
        16
        ——
       288
        48
        ——
12)768(64 feet.
        72
        ——
         48
```

RULE—Multiply the width by the thickness and that product by the length in feet and divide by 12.

CONTRACTORS' AND BUILDERS' ESTIMATE.

If a stick of timber be 10 inches square and 40 feet long, 10 x 10 equals 100 x 40 equals 4000 divided by 12 equals 333⅓ feet, or you can first find the contents of one side of timber or joists and multiply by the thickness the result will the same. Example— If a stick of timber be 24 feet long, 10 inches wide and 6 inches thick, 24 x 10 equals 240, divided by 12 equals 20 x 6 equals 120 feet.

Joists 2 1-2 by 8 inches 16 feet long.

```
  2)8
    2½
  ———
  2)20
   16½
  ———
   120
    20
    10
  ———
12)330(27½ feet,
   24
   ——
    90
    84
   ——
   6-12
```

or, say 16 1-2 feet long, 8 inches wide, 2 1-2 thick.

```
    16½
  2)  8
  ———
    128
      4
  ———
12)132(11
   12    2½
   ——   ——
   12   22
   12.   5½
   ——
        27½ feet.
```

Same as before.

To ascertain the cost of any number of feet of lumber; RULE, multiplipy the number of feet by the price paid per thousand; decimal points three places to the left.

Example, In the last joists estimated there is 27 1-2 feet, say at $14 per M.

```
   27½
  2)14
   ——
   108
    27
     7
   ——
  .385
```

Ans. Thirty-eight cents and 5 mills or thirty-eight and one-half cents.

Hay, Iron or anything of 2000 pounds to a ton; RULE, Multiply the number of pounds by the price paid per ton. and divide the product by 2; decimal point three places to the left.

Example, 884 lbs., of hay at $16 dollars per ton.

```
      884           Cost of 1068 lbs. at $22.50 per ton.
       16                   2)1068
      ---                    22½
     5304                   -----
      884                    2136
     ----                    2136
   2)14144                    534
    $7.072 Ans.            ------
                          2)24030
                           ------
                           12,015   Ans.
```

RULES AND DIRECTIONS FOR MEASURING WOOD AND BARK.

SUGGESTIONS TO SURVEYORS.

Examine the pile or load of wood carefully before taking the measurements, and if crooked sticks are in the pile, or the wood is crossed in piling, or in any way arranged as to make a larger pile than would be obtained by piling straight wood in as compact a manner as possible, make such deduction as in your judgment would make it sold measure of *well packed straight wood*.

If the pile of wood is in a curve, or part circle, always take the length on the inside of the circle.

Where wood is designed to be cut 4 feet in length, it should be that length from the point of one end to the commencement of the kerf of the other end. The same rule, where wood is cut any other length, is applicable, as regards those lengths.

In measuring a load of wood where chains are used to secure the stakes between the wood, a proper deduction from the height should be made for the space occupied by such chains.

In measuring a load of wood when the stakes are farther

apart at the top than at the bottom, the width of the load should be taken half way between the bottom and the top of the load.

In measuring a pile of wood which has a butment at either or both ends, deduct one foot from the length of each butment.

In measuring bark that is curled, or warped up, make such deduction as would make it a pile or load of straight bark.

RULE FOR SURVEYING WOOD BY A MEASURE OF FEET AND TENTHS OF FEET.

Multiply decimally the length, breadth and heighth together, and you will have the amount in feet and decimals of a foot which can be reduced to cords by dividing by 128.

SHORT RULE FOR DECIMAL MEASURE WHEN THE WOOD IS FOUR FEET IN LENGTH.

Multiply the length by the heighth and divide by 32 and you will have the amount in cords and fraction of a cord.

In measuring *unpacked stove* wood, deduct *one-third* of the *gross* measure to give solid measure,

RULE FOR SYRVEYING WOOD BY A MEASURE OF FEET AND INCHES.

Multiply duodecimally the length by the heighth and that product by the width and divide by 128 and you have the amount in cords and parts of a cord, as by the following example;

Find the solid contents of a pile of wood 9 ft. 5 in. long, 3 ft. and 7 in. high and 4 ft. 2 in. wide.

CONTRACTORS' AND BUILDERS' ESTIMATE.

```
            9—5'              12              128 ft.
            3—7'              12                12
          ─────────          ────            ──────
          5  5'  11"          144'             256
         28  3'                 7'             128
         ─────────           ────            ──────
         33  8'  11"          151'            1536'
             4   2'            12               12
         ─────────           ────            ──────
          5  7'  5"  10"'     302             3072
        134 11' 8"            151             1536
       ──────────             ────           ──────
  128)140  7' 1" 10"'(1 21766/221184 cord    18432"
      128                    1813"              12
      ────                     12             ──────
       12 ft.-7'-1"-10"'     ────             36864
   Reduced to a fraction of  3626             18432
          a cord.            1813            ───────
                             ────            221184"".
                             10"'
                             ────
                             21766"'
```

or, if solved by common fractions,

9 feet, 5 in. = 9 5-12 ft. = 113-12
3 " 7 " = 3 7-12 " = 43-12
4 " 2 " = 4 2-12 = 50-12

$\frac{113}{12} \times \frac{43}{12} = \frac{4859}{144} \times \frac{50}{12} = \frac{242450}{1728} = 140$ feet. $\frac{1030}{1728}$

= 1 cord 12 $\frac{1030}{1728}$ ft. or 1 $\frac{21766}{221184}$ cord as before.

or by reducing the whole to inches.

9 ft. 5 in.=113 inches,
3 " 7 " =43 "
4 " 2 " =50 "

113x43=4859x50=242450 inches reduced to cords, 242450 divided by 1728=140, $\frac{1030}{1728}$ ft. divided by 128=1 cord, 12 $\frac{1030}{1728}$ ft., or 1 $\frac{21766}{221184}$ cord as before.

NOTE.—It is usually considered of sufficient exactness to reject the parts of a foot if they are less than a half, or if they exceed half, to add a foot in which case, by the above example we have 1 13-128 cord.

To measure a load or pile of wood 12 ft. long and 4 ft. wide, call every 4 inches in height 1-8 of a cord. every 2 inches 1-16 and every inch 1-32.

To measure a load or pile of wood 8 ft. long and 4 ft. wide, call every inch in height 1-48 of a cord, every 2 in. 1-24, every 3 in. 1-16, every 4 inches 1-12 of a cord. Sixteen solid feet make 1 cord foot and a pile of wood 4 ft. long, 4 ft. wide and 1 foot high contains 1 cord foot and it is usually very convenient in measuring wood to take the amount in cord feet, which are reduced to cords by dividing by 8.

HIP RAFTERS.

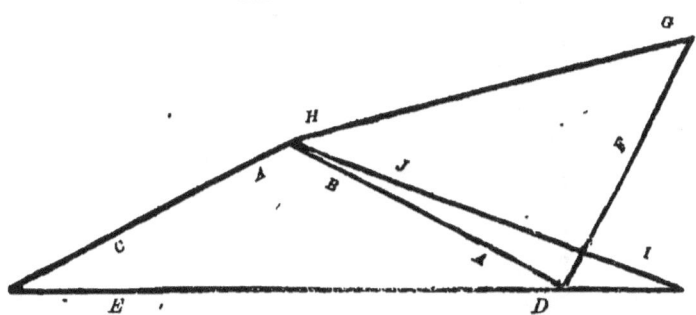

A simple and easy method to obtain the exact length and bevels of hip rafters for any pitch roof; (see diagram above,) *A B* and *A C* represents the main rafters, line *D E* the plate. The line *F* is drawn on a right angle, of main rafter *AB*. The line *F* must be in length of one-half the width of the building at point *G*.

The line from *G* to *H* gives the length of the hip-rafters required. The line *I J* is the same length as the line *G H*, the same as to let end *G* fall to *I* to top of plate *H*, remaining on its position at the peak of the roof. The line *I J* is to obtain the bevels at each end. Apply the same rule for any pitch roof. First draw the main rafter lines, then draw line *F* in a right angle with the pitch of the main rafter, in length of one-half the width of the building as described in the diagram, above from point G to *H*, will always give the length of any hip rafter of any pitch roof, with the greatest exactness.

EXAMPLE.

Suppose a building to be 24 feet wide, one-half the width is 12 feet, the rise of the main rafter is 6 inches to one foot; the length of the main rafter is 13 feet and 5 inches; in this case the right angle line *F* from main rafter will be 12 feet to point *G*; the line from *G* to *H* is 18 feet, the length of the hip-rafter required. The line *I J* also is 18 feet long, or the same line end *G* drops to the line of plate, as explained before, to give the bevels at each end.

OCTAGON.

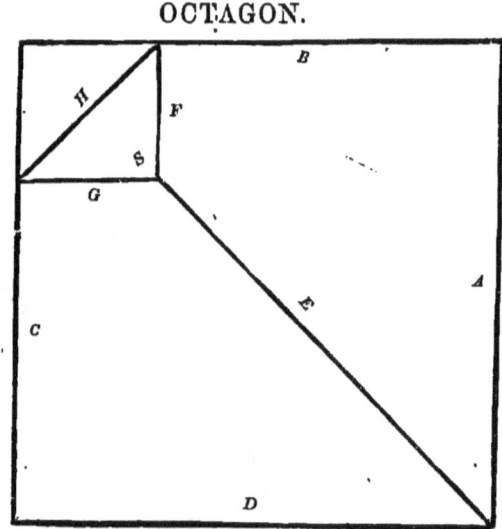

To obtain the equal sides of an octagon by a square. The plate above represents a building 50 feet sq. First make the square lines AB and CD; the line E. is drawn on a line from corner to corner of the square in length to correspond with the sides of the building which is 50 feet to point S.

To get the lines F and G, place the blade of the square on the outer lines at a point that the tongue of the square will strike the end of line E at S; then draw the line H, as shown in the plate. The line H will be the equal length of the eight square or sides of the octagon.

The same rule will apply to any size building or square to be octagoned, always remember that the line E must be in length to correspond with the width or sides of the building. If a building is 24 feet square the line E will be 24 feet long to point S; if 10 feet square, line E will be 10 feet. If a block is 10 inches square line E, will be 10 inches long to point S. The line H will always give the equal length of the sides.

The bevels of an Octagon. Place the blade of the square at 8 1-2 inches, the tongue at 20 1-2 inches will give the proper bevels.

PAINTING.

For estimating the cost of painting no general rule can be given. It is customary to charge for each coat of paint put on, and cevry part should be measured that is painted and allowances made for covering deep quirks in mouldings, carved surfaces, difficult mouldings, railings of iron &c

The condition of the surface to be painted governs the price per yard. It takes from 4 to 6 gallons of oil for 100 pounds of white lead, depending somewhat upon the heat or cold.

FRENCH OR MANSARD ROOF.

In estimating the amount of frame lumber contained in a building which requires a french or mansard roof. Add one-third to the amount of lumber contained in the pitch roof, that is the frame lumber above the plate. Take the number of feet of lumber contained in the rafters, collar girts and gable studs adding $\frac{1}{3}$ for a french roof. The diffrence cannot be stated exact as the pitch of the straight roof size of lumber &c. will vary but to assist in making an estimate of buildings of an ordinary size, this estimate will be practically correct.

PITCH OF ROOFS.

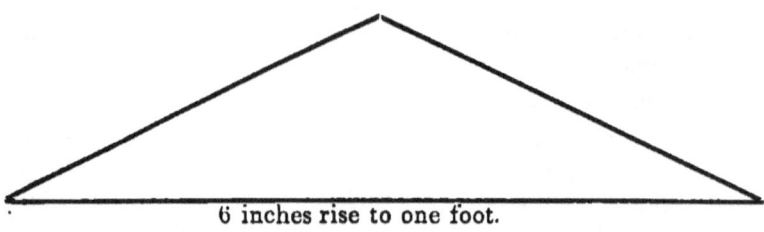
6 inches rise to one foot.

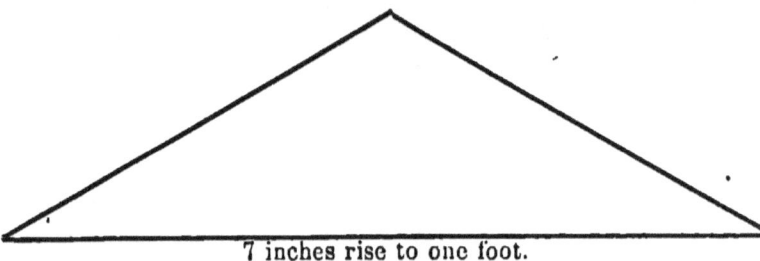
7 inches rise to one foot.

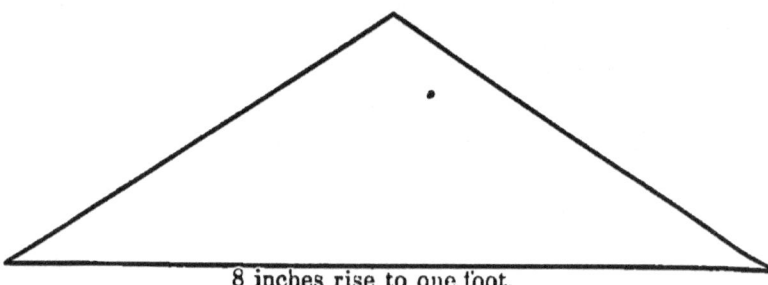
8 inches rise to one foot.

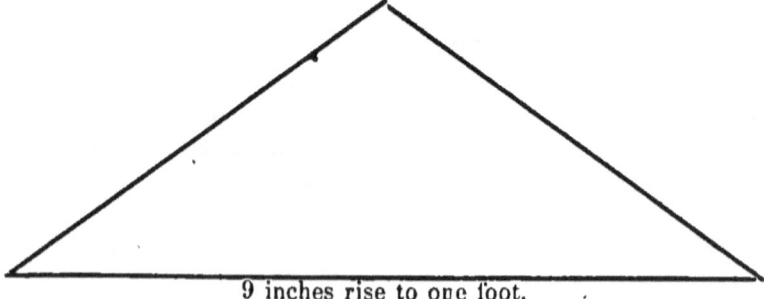
9 inches rise to one foot.

PITCH OF ROOFS.

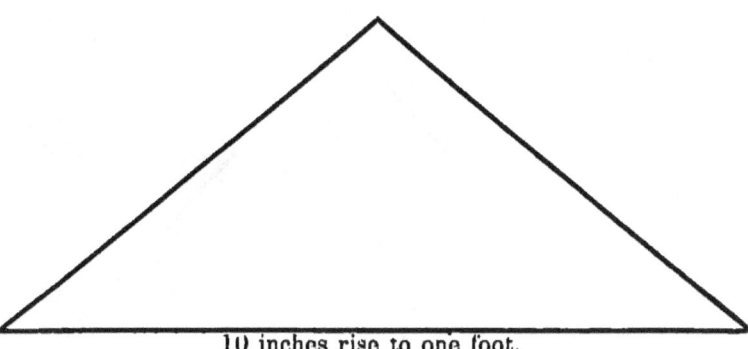
10 inches rise to one foot.

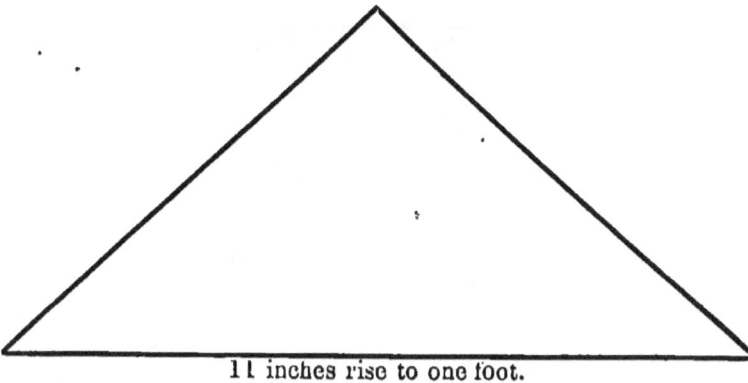
11 inches rise to one foot.

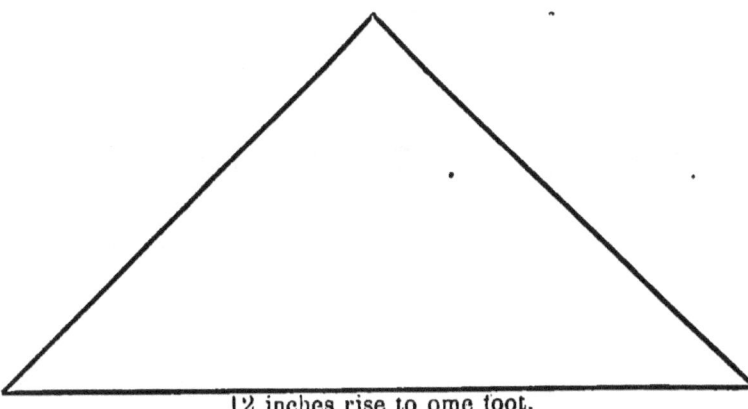
12 inches rise to one foot.

PITCH OF ROOFS.

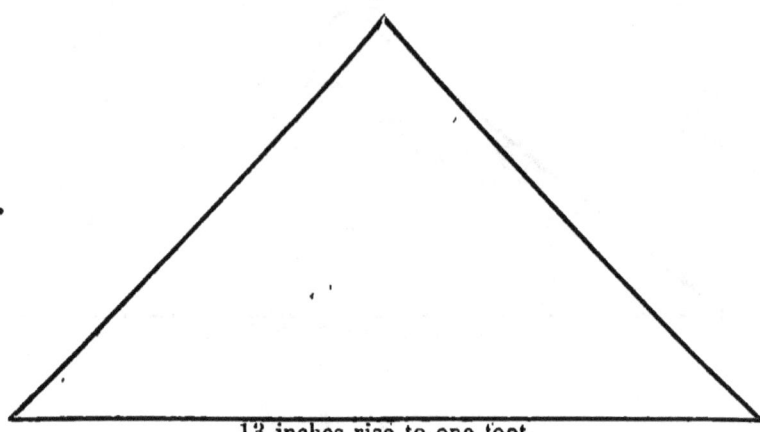

13 inches rise to one foot.

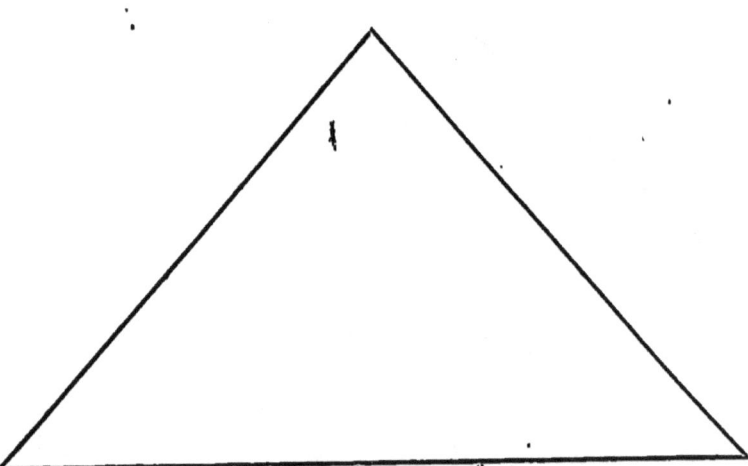

14 inches rise to one foot.

TABLE

Showing the number of pieces of square timber 1 inch long, from 1 inch to 3 inches square, in 1000 feet inch measure, and the number of feet, inch measure, in 1000 pieces of square timber 1 inch long, from 1 inch to 3 inches square.

USE OF TABLE.

Divide the number in the third column opposite the size required by the length in inches, and the quotient will be the number of pieces in 1000 feet.

EXAMPLE.—Find the number of pieces 10 inches long, $1\frac{1}{2}$ x $1\frac{1}{2}$ in 1000 feet. 64,000 divided by 10 equals 6,400 pieces.

Multiply the number in the fourth column opposite the size required by the length in inches, and the product will be the number of feet in 1000 pieces.

EXAMPLE.—Find the number of feet in 1000 pieces 10 inches long, $1\frac{1}{2}$ x $1\frac{1}{2}$; 15,625 multipled by 10, equals 156.25.

Length in Inches.	Size in Inches.	No. pieces in 1000 feet.	No. feet in 1000 pieces.
1	1 x 1	144000	6.944
1	$1\frac{1}{8}$ x $1\frac{1}{8}$	113777†	8.788
1	$1\frac{1}{4}$ x $1\frac{1}{4}$	92160	10.850
1	$1\frac{3}{8}$ x $1\frac{3}{8}$	76165†	13.129
1	$1\frac{1}{2}$ x $1\frac{1}{2}$	64000	15.625
1	$1\frac{5}{8}$ x $1\frac{5}{8}$	54532†	18.337
1	$1\frac{3}{4}$ x $1\frac{3}{4}$	47020†	21.267
1	$1\frac{7}{8}$ x $1\frac{7}{8}$	40900†	24.414
1	2 x 2	36000	27.777
1	$2\frac{1}{8}$ x $2\frac{1}{8}$	31888†	31.358
1	$2\frac{1}{4}$ x $2\frac{1}{4}$	28444†	35.156
1	$2\frac{3}{8}$ x $2\frac{3}{8}$	25229†	39.168
1	$2\frac{1}{2}$ x $2\frac{1}{2}$	23040	43.403
1	$2\frac{5}{8}$ x $2\frac{5}{8}$	20897†	47.851
1	$2\frac{3}{4}$ x $2\frac{3}{4}$	19041†	52.517
1	$2\frac{7}{8}$ x $2\frac{7}{8}$	17421†	57.400
1	3 x 3	16000	62.500

†Decimal.

Size of Glass. 4 light Win.	Size of Sash. Ft. In.	Ft. In.	Thickness			No. lbs. to each Weight. 1 3-8	1 1-2
10½ by 18	2 0⅝	by 3 5	1¼	..	1½		......
12 " 20	2 3⅝	" 3 9	1¼	1⅜	1½	3½	4
12 " 22	2 3⅝	" 4 1			1½		4
12 " 24	2 3⅝	" 4 5	1¼	1⅜	1½	4	4
13½ " 22	2 6⅝	" 4 1			1½		4½
13½ " 24	2 6⅝	" 4 5	1¼	1⅜	1½	4½	5
13½ " 26	2 6⅝	" 4 9	1¼	1⅜	1½	4½	5
13½ " 28	2 6⅝	" 5 1		1⅜	1½	5	5½
13½ " 30	2 6⅝	" 5 5			1½		5½
12½ " 32	2 6⅝	" 5 9			1½		6
13½ " 34	2 6⅝	" 6 1			1½		6½
15 " 24	2 9⅝	" 4 5			1½		5
15 " 26	2 9⅝	" 4 9			1½		5½
15 " 28	2 9⅝	" 5 1		1⅜	1½	5	5½
15 " 30	2 9⅝	" 5 5		1⅜	1½	5½	6
15 " 32	2 9⅝	" 5 9			1½		6½
15 " 34	2 9⅝	" 6 1			1½		7
15 " 36	2 9⅝	" 6 5			1½		7
16½ " 28	3 0⅝	" 5 1			1½		6
16½ " 30	3 0⅝	" 5 5			1½		6½
16½ " 32	3 0⅝	" 5 9			1½		6½
16½ " 31	3 0⅝	" 6 1			1½		7
16½ " 36	3 0⅝	" 6 5			1½		7½
18 " 28	3 3⅝	" 5 1			1½		6½
18 " 30	3 3⅝	" 5 5			1½		7
18 " 32	3 3⅝	" 5 9			1½		7½
18 " 34	3 3⅝	" 6 1			1½		8
18 " 36	3 3⅝	" 6 5			1½		8½
14 " 24	2 7⅝	" 4 5			1½		5
14 " 26	2 7⅝	" 4 9			1½		5½
14 " 28	2 7⅝	" 5 1			1½		6
14 " 30	2 7⅝	" 5 5			1½		6
16 " 28	2 11⅝	" 5 1			1½		6
16 " 30	2 11⅝	" 5 5			1½		6½
16 " 32	2 11⅝	" 5 9			1½		6½
16 " 34	2 11⅝	" 6 1			1½		7
16 " 36	2 11⅝	" 6 5			1½		7

Corresponding Sizes of 12 and 4-Lighted Windows.

7	by	9	same	size as	10½	by	18
8	"	10	"	"	12	"	20
8	"	11	"	"	12	"	22
8	"	12	"	"	12	"	24
9	"	11	"	"	13½	"	22
9	"	12	"	"	13½	"	24
9	"	13	"	"	13⅓	"	26
9	"	14	"	"	13½	"	28
9	"	15	"	"	13½	"	30
9	"	16	"	"	13½	"	32
10	"	12	"	"	15	"	24
10	"	13	"	"	15	"	26
10	"	14	"	"	15	"	28
10	"	15	"	"	15	"	30
10	"	16	"	"	15	"	32
10	"	17	"	"	15	"	34
10	"	18	"	"	15	"	36
11	"	14	"	"	16¼	"	28
11	"	15	"	"	16½	"	30
11	"	16	"	"	16½	"	32
11	"	17	"	"	16½	"	34
11	"	18	"	"	16½	"	36
12	"	14	"	"	18	"	28
12	"	15	"	"	18	"	30
12	"	16	"	"	18	"	32
12	"	17	"	"	18	"	34
12	"	18	"	"	18	"	36

To find size of Windows, add 3⅝ inches to width of glass, and 5 inches to heighth.

Blinds for Wooden Buildings, same width as Windows, and ½ inch longer. For Brick Buildings, same width, and 2 inches longer than windows.

MEASUREMENT OF LOGS
WITH THE USE OF CALIPERS.

CORD MEASURE.

TABLE 1—Gives the contents of logs in cord measure. This table is based upon the following rule viz: Square the diameter in inches, multiply by the length of the log in feet, divide the product by 144; result obtained in feet. Calipers applied in centre of log when of ordinary length. If logs are over 12 or 14 feet long the place of application of calipers should be moved towards large end of log. Logs 20 feet and over in length should be calipered in sections of 10 or 12 feet. In the use of this rule the diameter is usually taken over the bark. The place of application of calipers is however sometimes varied by contract between parties. This rule gives the entire contents of the log in cord measure without reference to its imperfections (always excepting rotten and worthless logs) or to what it may finally be sawed into. The chopper, the drawer, the buyer and seller receive full measure. The question of quality and discount to be considered when the price is fixed for the logs.

TABLE No. 1.

CORD MEASURE.

The figures indicating inches, are so many twelfth-parts of a foot.

MEASUREMENT OF LOGS,

WITH THE USE OF CALIPERS.

INCH BOARD MEASURE.

The object in this measurement, as claimed by the authors of tables, is to show the number of feet in one inch boards that may be obtained from a log of given dimensions.

In examining the many different rules and tables used for this purpose, I find myself confronted by figures as contradictory, as they are numerous. Selecting thirteen inch measure tables and rules, which have been used in New England, and taking as a basis, a log 12 feet long and 12 inches in diameter, I obtain the following results, viz:

Doyle's Rule	gives	48 feet.
Scribner's Table,	"	59 "
State Law of New Hampshire,	"	67 "
Younglove's Table,	"	68 "
Partridge's ⅞ Table, (reduced to inch)	"	70 "
Humphrey's Rule,	"	72 "
Wilson's Table,	"	75 "
Bangor Rule,	"	75 "
A pair of Calipers, (author unknown,	"	77 "
Holland's Rule,	"	78 "
Rule from Butt's Business Man's Assistant,	"	81 "
Cook's Table,	"	81 "
Derby's Table,	"	81 "

The entire contents of a log 12 feet long and 12 inches in diameter at centre, (or to be precise with a mean diameter of 12 inches,) is in inch measure, 113 feet.
Making the usual deduction of 25 per cent. for saw kerf, we have, 85 "
Deducting 20 per cent. (instead of 25) which approaches nearer mathematical accuracy, even though the saw make a cut of ¼ inch, we have 90 "

as the contents of the log in inch measure, making NO ALLOWANCE for slabs or straightening of edges.

In the above extracts, some of the tables which give the same measurement on a 12 inch log, differ widely on logs having a greater diameter. Such is the disagreement that it may be claimed that no two authors have arrived at the same results. In the brief space at our disposal it will be impossible to explain in detail the cause of these variations. I can only speak of the more important.

Some tables give the contents of the log when sawed into boards all square edge, with parallel sides. By this rule the diameter is taken at the small end of the log; in which case all the swell of the log and also the contents of the slabs, however thick, is not included in the measurement.

Another author works from the same basis, but allows a certain number of wany edge boards to be sawed; which number varies according to the size of the log. Others allow the boards to be sawed tapering, to conform to the shape of the tree, squaring the edges of most of the boards. Another makes the boards all round edge.

Some authors openly avow that they make allowance for " CONCEALED DEFECTS," &c., &c.

Now while the authors of these tables differ thus widely theoretically, in reference to the manner in which these same inch boards should be sawed, still other elements come in, in practice, which tend to add additional complications.

First of all, the timber itself has much to do with its production, especially when sawed into boards. The size, age, shape and kind, all affected greatly by the nature of the soil on which it grows, produces ever varying results. It is not an easy matter to find two sawyers who will obtain the same number of feet of boards from a given quantity of logs. A. saws his boards so thick and gives such width in measuring that they will hold out measure when thoroughly seasoned. B. finds it difficult to set his guage so that the boards ever come out *quite one inch thick* even before they are seasoned. C, (representing no small number,) meets with many difficulties, and his boards frequently leave the saw in the form of clapboards, or shingles.

All of these sawyers (especially the latter.) will talk knowingly of measurements.

If the authors of these tables were dealing with the question of the measurement or solidity of the log itself, they should come to the same conclusions; for only one mathematical rule could be used in the solution. But these inch "measure tables" so called, seek not the measurement of the log, but a measurement of the product of the log in inch boards.

With equal propriety tables might be made giving the supposed contents of logs when manufactured into shingles, clapboards, joist, pail staves, chair stretchers, broom handles, &c.; all of which would be convenient for the manufacturer as tables of reference, but could not consistently be used for the purposes of general measurement. Whatever might have been the fact thirty or forty years ago, it is unquestionably true today, that not one log in ten that is measured by these same tables, is ever sawed into one inch boards; hence in their use there is at least a misapplication.

It makes but little difference which table or method of measurement is adopted; you may use Doyle's rule, giving 48 feet in a 12 inch log, or Derby's giving 81 feet; no injustice is done if the facts are all made known to the parties interested.

The man who draws the logs at a stipulated price, with the understanding that a table is to be used giving him 81 feet, would not be willing to draw them at the same rate if they were to be computed by a table giving him only 48 feet. The purchaser in giving $10 per M for logs by Doyle's rule, would not be satisfied to pay the same price if they were measured by Derby's.

While the purchasers of logs have generally some knowledge of tables, and may be trusted to guard their own interests; it is not unreasonable to assert that not one man in a hundred engaged in lumbering (including the laborer,) has any knowledge of these said inch measure tables.

In exposing the inconsistencies of these tables, I desire to throw no obstacles in the way of the purchaser of lumber, tending to lessen his legitimate profits. I have, however, endeavored to present a few facts, hoping that all persons interested in lumbering may be induced, to either accept "cord measure" as heretofore explained, as the *standard of measurement* or some inch measure table, which is based upon the solidity of the log and which gives its contents.

It must be evident to all that for the *purposes of measurement* alone, the proper question for consideration is, the number of feet and inches a log contains of sound timber; all other matters should be settled in the adjustment of prices.

www.ingramcontent.com/pod-product-compliance
Lightning Source LLC
Chambersburg PA
CBHW030340170426
43202CB00010B/1188